The Inorganic Chemistry of the Non-metals

Methuen Studies in Science

GENERAL EDITOR   J. M. Gregory MA., D.Phil., Winchester College

CONSULTANT EDITORS   B. E. Dawson B.Sc., Ph.D., King's College, London

R. Gliddon B.Sc., Ph.D., Clifton College, Bristol

This series provides students with concise, introductory surveys of important topics in the physical, chemical and biological sciences. The series is designed to assist sixth form students preparing for entry to university or college, and to meet the needs of university students preparing for more advanced studies.

Chemical Equilibrium	J. S. Coe
Enzymes	Alan D. B. Malcolm
Energy in Chemistry	B. E. Dawson
Alternating Currents	J. M. Gregory
Nature Conservation	W. M. M. Baron
Logical Control Systems	Keith Morphew
Aspects of Isomerism	Peter Uzzell
The Inorganic Chemistry of the Non-metals	John Emsley
Atomic and Molecular Weight Determination	R. B. Moyes
The Mechanical Properties of Materials	R. A. Farrar

Forthcoming

Oscillations	Ian B. Hopley
The States of Matter	A. Ralph Morgan
Transition Metal Chemistry	Jeff Thompson
Crystals and their Structures	I. F. Roberts
Catalysis in Chemistry	A. J. B. Robertson
Molecular Spectroscopy	Aline Bradshaw
Kinetics and Mechanisms of Reactions	B. E. Dawson
Waves	Ian B. Hopley
Radioisotopes	D. J. Hornsey
Fundamental Electrostatics	S. W. Hockey

Methuen Studies in Science

The Inorganic Chemistry of the Non-metals

JOHN EMSLEY Ph.D.

*Lecturer in Chemistry,
King's College, London*

Methuen Educational Ltd
LONDON · TORONTO · SYDNEY · WELLINGTON

First published 1971
by Methuen Educational Ltd
11 New Fetter Lane, London EC4
© 1971 by John Emsley
Printed in Great Britain by
William Clowes & Sons Ltd.,
London, Colchester and Beccles

SBN 423 87320 2   non-net
        423 86120 4   net

All rights reserved.
No part of this publication
may be reproduced, stored in
a retrieval system, or transmitted
in any form or by any means,
electronic, mechanical, photocopying,
recording or otherwise, without the
prior permission of the publisher.

Distributed in the U.S.A. by
Barnes and Noble, Inc.

Contents

Preface

1	The Non-metals and the Periodic Table	1
2	The Covalent Bond	5
3	The Structure of Molecules	16
4	Hydrogen	20
5	Boron	26
6	Carbon and Silicon	31
7	Nitrogen and Phosphorus	38
8	Oxygen, Sulphur, and Selenium	45
9	Fluorine, Chlorine, Bromine, and Iodine	51
10	The Rare Gases	55
Bibliography		58
Index		59

Preface

In the past twenty years there has been a renaissance of inorganic chemistry. This has led to the discovery of a large number of compounds and many of these have seemingly been at odds with the old laws of valency. The xenon fluorides and the boranes are examples of such compounds among the non-metals.

The upsurge in inorganic chemistry has coincided with the phenomenal growth of spectroscopy in recent years. Not only are more new compounds being discovered but more is being learned of their structures, their bonding, and their energetics. The old questions once asked of a particular compound were: How is it prepared? What are its physical properties? How does it react with common reagents? These questions are still worth asking of course but other, more basic questions now frame themselves. Why does a particular compound exist at all? Why does it have a certain structure? Why does it react as it does? These are the newer type of question in which the *how* of factual information has given way to the *why* of explanation. In this monograph I have tried to show how this change is affecting the inorganic chemistry of the non-metals.

At appropriate intervals in the text there are outlines of simple experiments which demonstrate a point being made in the text. For suggestions about many of these I should like to thank Dr B. E. Dawson. Questions are similarly dispersed throughout the book and these are designed to encourage the reader to deduce for himself the reasons for reported observations. Accompanying the chapters on the non-metal elements are tables of bond energies and bond lengths. Again the interested reader is encouraged to make use of these as a fresh basis for comparing and predicting the chemistry of the non-metals and their compounds.

King's College, London　　　　　　　　　　　　　　　　　　　　　　　　　　　　J. E.

Table 1 The periodic table of the elements

1s block

H	He
1	2

s block

Li	Be
3	4
Na	Mg
11	12
K	Ca
19	20
Rb	Sr
37	38
Cs	Ba
55	56
Fr	Ra
87	88

d block

Sc	Ti	V	Cr	Mn	Fe	Co	Ni	Cu	Zn
21	22	23	24	25	26	27	28	29	30
Y	Zr	Nb	Mo	Tc	Ru	Rh	Pd	Ag	Cd
39	40	41	42	43	44	45	46	47	48
*	Hf	Ta	W	Re	Os	Ir	Pt	Au	Hg
57-71	72	73	74	75	76	77	78	79	80
†									
89									

p block

B	C	N	O	F	Ne
5	6	7	8	9	10
Al	Si	P	S	Cl	Ar
13	14	15	16	17	18
Ga	Ge	As	Se	Br	Kr
31	32	33	34	35	36
In	Sn	Sb	Te	I	Xe
49	50	51	52	53	54
Tl	Pb	Bi	Po	At	Rn
81	82	83	84	85	86

f block

*	La	Ce	Pr	Nd	Pm	Sm	Eu	Gd	Tb	Dy	Ho	Er	Tm	Yb	Lu
	57	58	59	60	61	62	63	64	65	66	67	68	69	70	71
†	Ac	Th	Pa	U	Np	Pu	Am	Cm	Bk	Cf	Es	Fm	Md	No	Lw
	89	90	91	92	93	94	95	96	97	98	99	100	101	102	103

1 The non-metals and the periodic table

There are twenty non-metal elements. In the periodic table these consist of the 1s-block and the top right-hand half of the p-block of the elements as shown in Table 1. The division of the elements into metals and non-metals is not as simple as it appears, because there is no single property which can be used to classify all the elements as either metals or non-metals. Some elements behave in some instances as metals and in others as non-metals, and these are generally given a class to themselves and called metalloids. Table 2 shows only the s- and p-block elements (together called the main group elements) with their electronic configurations. The boundary between those elements which are metals by any definition and those which are non-metals by any definition is shown in Table 2 and the metalloids can be seen to straddle this boundary.

Table 2   The main group elements and their electronic configurations

n	M1 ns^1	M2 ns^2	M3 ns^2np^1	M4 ns^2np^2	M5 ns^2np^3	M6 ns^2np^4	M7 $1s^1$ H ns^2np^5	M8 $1s^2$ He ns^2np^6
2	Li	Be	B	C	N	O	F	Ne
3	Na	Mg	Al	Si	P	S	Cl	Ar
4	K	Ca	Ga	Ge	As	Se	Br	Kr
5	Rb	Sr	In	Sn	Sb	Te	I	Xe
6	Cs	Ba	Tl	Pb	Bi	Po	At	Rn
7	Fr	Ra						
	Metals			Metalloids			Non-metals	

The chemistry of two of the non-metals is very restricted. These are astatine and radon, which are the non-metals at the bottom of groups M7 and M8. These are both radioactive in all their isotopes which are mostly short-lived. Their chemistries have only been won as a result of patient and careful research using special apparatus for dealing with amounts of material often too small to be visible with the unaided eye. As far as is known they resemble very closely the elements above them, that is iodine and xenon.

The most important non-metal is carbon. Together with all its compounds it makes up the largest branch of chemical knowledge — organic chemistry. The reason for this lies in its ability to form strong carbon-to-carbon bonds and this leads to the formation of a vast number of molecules ranging from the very simple to the exceedingly complex. Some of the simpler carbon compounds with only one carbon atom fall within the province of inorganic chemistry but these form only a very small portion of the chemistry of carbon.

Table 3 Melting points, boiling points, and electrical conductances of the main group elements

M1	M2	M3	M4	M5	M6	M7	M8
						H	He
					m.p./(K)	14	3
					b.p./(K)	20	4
		Electrical conductance at 273 K/(ohm^{-1} cm^{-1})				—	—
Li	Be	B	C	N	O	F	Ne
454	1556	2300	3800*	63	54	53	24
1604	*2750*	*4200*		*77*	*90*	*85*	*27*
0·117	**0·169**	**<0·001**	**<0·001**	—	—	—	—
Na	Mg	Al	Si	P	S	Cl	Ar
371	923	933	1696	317	392	172	84
1163	*1393*	*2720*	*2953*	*553*	*718*	*239*	*87*
0·238	**0·224**	**0·377**	**<0·001**	**0**	**0**	—	—
K	Ca	Ga	Ge	As	Se	Br	Kr
336	1123	303	1033	886*	490	266	116
1039	*1765*	*2510*	*3103*		*961*	*331*	*120*
0·163	**0·292**	**0·019**	**<0·001**	**0·029**	**0**	—	—
Rb	Sr	In	Sn	Sb	Te	I	Xe
312	1043	429	505	904	723	387	161
974	*1643*	*2320*	*2960*	*1910*	*1360*	*457*	*165*
0·080	**0·043**	**0·119**	**0·087**	**0·026**	**<0·001**	**0**	—
Cs	Ba	Tl	Pb	Bi	Po	At	Rn
302	983	577	600	544	527	—	202
958	*1911*	*1713*	*2024*	*1832*	*1235*	*653*	*211*
0·053	**0·016**	**0·011**	**0·048**	**0·009**	—	—	—

* sublimes

Having indicated the elements to be included in the term 'non-metals' we can see what these have in common and how they differ from the metals. The melting points (m.p.), boiling points (b.p.), and electrical conductances of the elements themselves are the kinds of physical property which distinguish metals and non-metals. In Table 3 the values of these properties are shown for the main group elements. The reason why these properties reflect the differences is that they are closely related to the type of bonding which exists in the elemental state. Metals are held together by strong bonds (hence the high m.p.'s and b.p.'s) in which the bonding electrons have a characteristic freedom of movement through the body of the metal (hence the high conductances). Table 3 also shows boron, carbon, and silicon to have high m.p.'s. Again this is due to strong bonding, but in these elements the bonds are covalent in

Table 4 Melting points and boiling points of the fluorides of the main group elements

M1	M2	M3	M4	M5	M6	M7	M8
						HF	He
						m.p./(K) 190	–
						b.p./(K) 293	–
LiF	BeF_2	BF_3	CF_4	NF_3	OF_2	F_2	Ne
1118	1070*	144	89	64	49	53	–
1954		264	91	144	128	85	–
NaF	MgF_2	AlF_3	SiF_4	PF_3	SF_6	ClF	Ar
1268	1536	1530*	177*	121	209*	117	–
1977	2500			172		173	–
KF	CaF_2	GaF_3	GeF_4	AsF_3	SeF_6	BrF_5	KrF_2
1129	1693	1223*	247*	267	226*	210	230*
1775	2773			336		233	
RbF	SrF_2	InF_3	SnF_4	SbF_3	TeF_6	IF_5	XeF_2
1048	1673	1443	978*	563	234*	282	413
1681	2733	1473		592		371	–
CsF	BaF_2	TlF_3	PbF_2	BiF_3	PoF_2	AtF	RnF_2
955	1593	823†	1095	1000	–	–	–
1300	2533		1563	**	–	–	–

* sublimes
† disproportionates to TlF and F_2
** not reported

nature and the electrons are *localized* and lack mobility with the result that electrical conductances are much lower than found in metals. Table 3 shows changes in properties down the groups and along the rows of the periodic table. In some cases the changes are unusual but are capable of explanation. For example, gallium, indium, and thallium have much lower m.p.'s than the elements preceding them, whereas the m.p. of aluminium is very like that of the element preceding it, magnesium. (Why?)

Not only the properties of the elements but those of their compounds reflect the metal and non-metal boundary. The m.p.'s and b.p.'s of the typical fluoride compounds are listed in Table 4, fluorine being chosen because this element forms compounds with all but three of the elements. A clear division can be seen between the metals which form ionic fluorides with strong electrostatic forces holding the solid together and the non-metals which form covalent molecules which are held together in the solid state by much weaker forces called van der Waal's forces.

The change from metal to non-metal along a row of the periodic table can be followed by studying the chlorides of the elements of the third row — $NaCl$, $MgCl_2$, Al_2Cl_6, $SiCl_4$, PCl_3 or PCl_5, S_2Cl_2 and Cl_2. A set of simple experiments can be carried out to investigate the effect of heating these chlorides, and in certain cases the b.p.'s can be measured. The solubility or reaction with water can also be used as a basis of classification.

One feature of the chemistry of the non-metals which stands out more than any other is the type of chemical bond which these elements generally form. This is the covalent bond.

2 The covalent bond

The covalent bond, like all types of chemical bond, is easy to describe but hard to explain. Covalent bonds are the bonds which hold the atoms in molecules together, and the ideas behind the terms *covalent bond* and *molecule* are inseparable. The other two kinds of strong chemical bond are the ionic and the metallic. In these cases it is not meaningful to talk of molecules, because each ion or metal atom is bonded to all surrounding ions or metal atoms.

In a molecule the number of covalent bonds is related to the number of atoms in the molecule. If there are n atoms then there will be $n-1$ covalent bonds (or n covalent bonds if the molecule has a single ring structure). For example, phosphorus oxochloride, $POCl_3$, has five atoms and therefore four covalent bonds, benzene, C_6H_6, has twelve atoms and is cyclic so has twelve covalent bonds. Although the number of bonds in a molecule is fixed, the number of bonds a particular atom forms depends upon its valency, which in turn depends upon the number of unpaired electrons which that atom brings to the molecule. The key to covalent bond formation is the pairing of electrons. The most common kind of covalent bond is one that is held together by an electron-pair in which each atom supplies one electron.

The simplest covalent bond is that of the hydrogen molecule, H_2. This molecule contains only two electrons, one from each hydrogen atom. These electrons pair off and spend most of their time in the space between the two atoms. By so-doing they reduce the electrostatic repulsion between the two nuclei, which are then able to approach each other more closely until a position is reached in which the coulombic forces of repulsion and attraction are balanced. Such a pair of electrons is called a bonding pair and this type of covalent bond is called a sigma bond (σ-bond). This is illustrated by Fig. 1 where the potential energy of a diatomic molecule AB is plotted as a function of the distance between the two nuclei. As the two atoms approach, the energy of the system falls to a minimum corresponding to the formation of the covalent bond. If the atoms approach still nearer, the forces of repulsion outweigh those of attraction and there is a rapid increase in the potential energy of the system.

In some molecules the formation of the covalent bond may involve the pairing of four electrons into two bonding pairs. One of these pairs is the strong σ-bond lying between the two atoms, the other pair forms a weaker bond partly 'surrounding' the σ-bond and this is called a pi bond (π-bond). Together these constitute what is referred to as a double bond. There is also the much rarer triple bond which has six electrons associated with it in the form of three bonding electron-pairs. These are a σ-bond and two π-bonds which 'surround' the σ-bond. A single bond is symbolized by a single line drawn between the formulae of the two atoms composing the bond, e.g. H–F; a double bond by two parallel lines, e.g. O=C=O; and a triple bond by three parallel lines, e.g. N≡N.

It is not necessary for each atom of a bond to contribute equal numbers of electrons. In some single bonds one of the atoms supplies both of the electrons but the result is still a σ-bond. These bonds are sometimes given the names donor or dative bonds but there is basically no difference between such bonds and the bonds in which each atom contributes its own electron. If it is wished to emphasize this type of bond an arrow sign or charge transfer is used. For instance, in the adduct formed by ammonia and boron trifluoride the bonding electron-pair is supplied by the

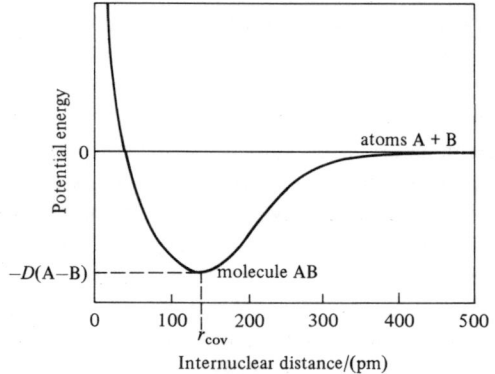

Fig. 1. *Potential energy of a diatomic molecule*

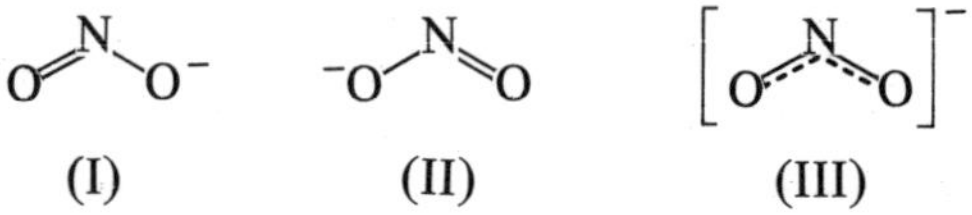

Fig. 2. *Bonding in the nitrite ion*

ammonia molecule, and a nitrogen-boron bond is formed. This can be signified by writing the formula of the adduct as $H_3N \rightarrow BF_3$ or $H_3\overset{+}{N}-BF_3$, but generally it is written $H_3N \cdot BF_3$.

In some molecules one finds the phenomenon of delocalized π-bonding, also called conjugate bonding or resonance. In its simplest form this occurs when an atom is connected by a single and a double bond to two atoms of the same sort. The nitrite ion, NO_2^-, is a good example (Fig. 2). This can be written as (I) or (II) and the π-bond is said to be localized. But we know this does not represent the true state of affairs because, if it did, we should be able to distinguish the two bonds, since single bonds are longer than double bonds. Both bonds in the nitrite ion are the same length and are equivalent in every respect. Each N—O bond has a certain proportion of double-bond character and the π-bonding is spread over all the atoms. The π-bonding is said to be *delocalized*. A better way of writing the structure of NO_2^- is (III), in which the basic σ-bonds are shown by lines and the delocalized π-bond by a dotted line linking those atoms over which it is delocalized.

Delocalization is not limited to two bonds. In the benzene molecule it embraces all the carbon atoms of the ring. In the first place these are held together by σ-bonds, as Fig. 3(IV) shows. In addition to the six electron-pairs forming the σ-bonds there are a further three electron-pairs for π-bond formation. These are not localized as in (Va), although commonly written as such in the abbreviated formula (Vb). Instead they are completely delocalized over all the carbon atoms of the ring (VIa), and the symbol (VIb) is often used as a more accurate representation of benzene. In delocalized π-bonding systems, all the delocalized bonds are the same length and this is true of benzene.

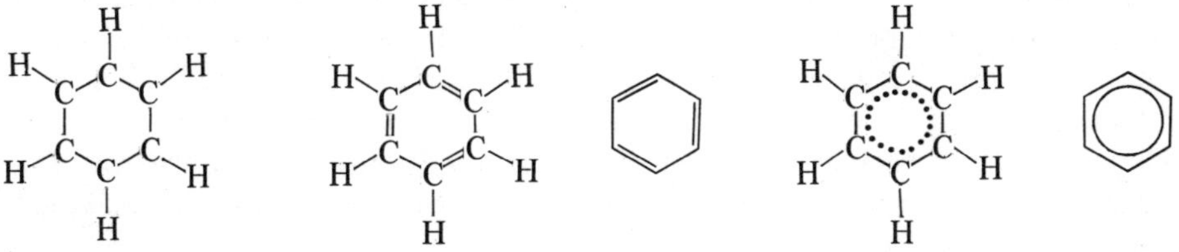

Fig. 3. Bonding in benzene

The carbonate, CO_3^{2-}, nitrate, NO_3^-, sulphate, SO_4^{2-}, acetate, $CH_3CO_2^-$, and many other oxoanions have delocalized π-bonding systems. With such ions it is possible to draw formulae with double and single bonds and this is often done to show the negative charge on such ions, but it should always be remembered that this is done for convenience at the expense of accuracy.

Properties of covalent bonds

The properties of a bond which interest a chemist are its length, strength, and the position of the bonding electron-pairs. The length of a bond is the distance between the nuclei of the two atoms making the bond, and this distance can be measured very precisely by a number of different methods. The strength of a bond is the energy needed to break it, and this is also the energy which is released when the bond is formed. In the case of diatomic molecules this energy can be measured very accurately, but with molecules containing more than two atoms only approximate values of this energy can be obtained. The position of the bonding electron-pairs cannot be specified with any degree of certainty. Yet it is known that some atoms attract electrons more strongly than others and the term 'electronegativity' is used to describe the ability of an atom in a molecule to attract electrons. Electronegativity cannot be measured directly so it has to be derived from those properties of molecules which reflect the electronegativity of its atoms. We shall look at each of these properties of covalent bonds more closely.

Bond length (r) and covalent radius (r_{cov})

It is possible to measure the distance between the nuclei of the atoms of a molecule by means of X-ray diffraction, neutron diffraction, and microwave spectroscopy techniques. Figure 1 shows what is meant by bond length, which is the distance between two atoms forming a covalent bond when the potential energy of the system is at a minimum. Bond lengths are of the order of 10^{-10} m, but are usually reported in nm or pm units, the latter being preferred. The shortest covalent bond is that of H_2 which is 75 pm (or 0·075 nm). The normal range of covalent bonds is 100 to 200 pm with only a few longer than this; for example, I_2 has a bond length of 267 pm.

Covalent bonds are generally shorter than either ionic bonds or metallic bonds. In sodium chloride the Na^+ and Cl^- ions are 281 pm apart. In iron the Fe atoms are 248 pm apart and this is exceptional in being the shortest internuclear distance of any metal; in magnesium, which is perhaps more typical, the Mg atoms are 320 pm apart.

Although some bonds have been measured to a high degree of precision, e.g. the C–C distance in diamond is 154·452 ± 0·014 pm, the chemist is normally satisfied with values reported to the nearest pm. From the large number of molecules whose bonds have been measured it has been observed that the bond length between a given pair of atoms is often the same in different types of molecules. Thus r(O–H) is 96 pm in water, methanol, and formic acid, and 97 pm in hydrogen peroxide. The bond length between a particular pair of atoms can vary if the nature of the bonding changes. Single bonds are longer than double bonds, which in turn are longer than triple bonds. For instance, the C–C bond length in ethane is 154 pm (single bond), in ethylene is 134 pm (double bond), and in acetylene is 120 pm (triple bond). The same sort of variations occur with the other elements capable of multiple bonding as Tables 12, 14, and 16 show. Delocalized bonds have bond lengths between the values for a single and a double bond. In benzene, r(C–C) is 139 pm which is below the average for a single and a double bond, 144 pm. The reader is invited to speculate on the reason for this and what it means in terms of stability of the benzene ring. Bond lengths can be used as a guide to the type of covalent bonding in a molecule.

We have already seen that to a first approximation the length of a particular bond is independent of its environment. This would follow if atoms were a definite size which did not change on forming a bond. If atoms are imagined to be spheres then it should be possible to deduce their radius from bond lengths. If a bond is formed by two atoms of the same type, then the bond length should be twice the radius of a single atom and this is called the covalent radius and given the symbol r_{cov}. Strictly speaking this is the single covalent bond radius and, although it is possible to speak of double and triple covalent bond radii, they are relatively unimportant because of the small number of elements which form such bonds. For carbon, r(C–C) is 154 pm and therefore r_{cov} is 77 pm for carbon. For chlorine, r(Cl–Cl) is 199 pm giving r_{cov}(Cl) a value of 100 pm. If our assumption that bond lengths are the sum of the covalent bond radii of the atoms of the bond then r(C–Cl) will be r_{cov}(C) + r_{cov}(Cl) which is 177 pm. In CCl_4 and most other compounds with a C–Cl bond the bond length is about 176 pm, which confirms us in our original assumption. The single bond covalent radii of the non-metals are given in Table 5.

It has already been mentioned that the bond length of the hydrogen molecule is 75 pm, from which one would calculate a covalent radius of 38 pm for hydrogen. This is not the value given in Table 5. If 38 pm were to be used in the estimation of bond lengths these would be found to be always too long. If r_{cov}(H) is calculated from such compounds as HCl, CH_4, etc., a value of about 30 pm is obtained and this can be used to predict successfully the length of bonds to hydrogen. The bond in the hydrogen molecule itself does not appear to be typical of hydrogen in general.

Table 5 The single covalent bond radii of the non-metals, r_{cov}/(pm)

B	C	N	O	F	
			H 30		
c. 70	77	71	66	64	
	Si 117	P 109	S 104	Cl 100	
			Se 114	Br 111	Kr 116
				I 128	Xe 121

Boron is another atom for which r_{cov} varies somewhat. The reader should calculate r_{cov}(B) from the bond lengths of Table 11 to demonstrate this. Another useful exercise is to draw up tables of double and triple covalent radii for the second-row elements and this can be done using the data from the tables of bond energies and bond lengths which accompany each of the chapters dealing with specific non-metal elements in this book.

Bond energy (E) and bond dissociation energy (D)

The bond dissociation energy of a diatomic molecule is the energy required to break the molecule into two quite separate atoms:

$$A–B \rightarrow A + B$$

This energy can be measured experimentally. In the case of iodine it can be determined spectroscopically by using the apparatus shown in Fig. 4. In this experiment a sealed

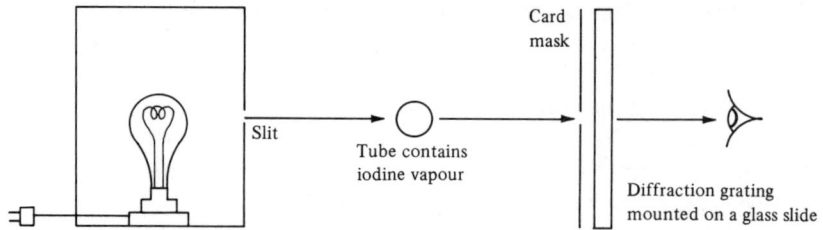

Fig. 4. *The bond dissociation energy of iodine*

tube containing iodine vapour is illuminated along its length by a slit of light from a 100 W bulb, and the spectrum is viewed through a diffraction grating mounted on a glass slide. The first position of continuous absorption represents the onset of dissociation of the iodine molecule. The wavelength, λ, of the light at this point can be used to calculate its energy, E, from Planck's relationship:

$$E = \frac{hc}{\lambda}$$

where h is Planck's constant = $6 \cdot 626 \times 10^{-34}$ J s, and c is the velocity of light = $2 \cdot 978 \times 10^8$ m s^{-1}. In this experiment the iodine molecule dissociates into two iodine atoms, one of which is in an excited state, having 90 kJ mol^{-1} extra energy. This amount must therefore be subtracted from E calculated from λ to give the dissociation energy of iodine, $D(I_2)$.

The bond dissociation energies of the more common diatomic molecules are given in Table 6. With molecules having more than one bond it is still possible to speak of the bond dissociation energy of a particular bond. For example the energy required to break one of the bonds of the water molecule has been measured:

$$H_2O \rightarrow HO + H$$

and is 500 kJ mol^{-1}. But for the vast majority of molecules it is not possible to measure the bond dissociation energy of individual bonds. Instead one uses *mean bond energies*, often simply called bond energies and given the symbol E.

When a molecule is shattered into atoms, and all the bonds are broken, then the energy released is called the *heat of atomization*, $\Delta H^{\ominus}_{atom}$. This is the sum of all the bond dissociation energies of the molecule. It is possible to calculate $\Delta H^{\ominus}_{atom}$ for a molecule, say AB_n, from the heats of formation of the molecule and the atoms of which it is composed, using the relationship:

$$\Delta H^{\ominus}_{atom}(AB_n) = \Delta H^{\ominus}_f(A) + n\Delta H^{\ominus}_f(B) - \Delta H^{\ominus}_f(AB_n) \qquad (1)$$

For a simple molecule AB_n in which there is only one type of bond A—B, the mean bond energy is $\Delta H^\ominus_{atom}(AB_n)/n$ and this is denoted $E(A-B)$.

In the case of ammonia, NH_3, the value of $E(N-H)$ is calculated as follows:

$$\begin{aligned}\Delta H^\ominus_{atom}(NH_3) &= \Delta H^\ominus_f(N) + 3\Delta H^\ominus_f(H) - \Delta H^\ominus_f(NH_3) \\ &= (+475) + 3(+218) - (-46) \\ &= 1175 \text{ kJ mol}^{-1}\end{aligned}$$

This represents the sum of the energies of the three N–H bonds, therefore the mean bond energy, $E(N-H)$, is one-third of this, i.e. 392 kJ mol^{-1}.

With molecules which have more than one type of bond a slightly different approach is necessary. Again $\Delta H^\ominus_{atom}$ is used as the basis of bond energies but it cannot directly be divided among the bonds because there are now two unknowns and only one equation. To get round this difficulty it is necessary to assume that the bond energy of a bond is the same in all molecules, and to use bond energies already known to calculate unknown bond energies. The fact that the bond length of a bond

Table 6 Bond dissociation energies, D, of common diatomic molecules

Molecule	D/(kJ mol^{-1})	Molecule	D/(kJ mol^{-1})	Molecule	D/(kJ mol^{-1})
H_2	435	O_2	410	HCl	425
F_2	153	N_2	950	HBr	365
Cl_2	243	NO	628	HI	298
Br_2	196	CO	1074	ClF	253
I_2	152	HF	562	BrCl	219

tends to be the same in different molecules supports this assumption, since bond length and bond energy are related to each other.

The problem is illustrated by diboron tetrachloride, B_2Cl_4, which has the structure shown. To calculate $E(B-B)$ and $E(B-Cl)$ we need to know the heats of atomization

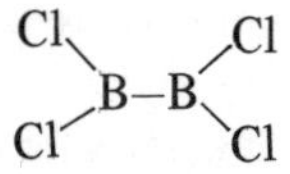

$$\begin{array}{cc} Cl & Cl \\ \diagdown & \diagup \\ & B-B \\ \diagup & \diagdown \\ Cl & Cl \end{array}$$

of both BCl_3 and B_2Cl_4. These are related to the bond energies as follows:

$$\Delta H^\ominus_{atom}(BCl_3) = 1329 \text{ kJ mol}^{-1} = 3E(B-Cl) \qquad (2)$$

$$\Delta H^\ominus_{atom}(B_2Cl_4) = 2110 \text{ kJ mol}^{-1} = E(B-B) + 4E(B-Cl) \qquad (3)$$

From (2) we see that $E(B-Cl)$ is 422 kJ mol^{-1} and this value can be substituted in (3) to give 342 kJ mol^{-1} for $E(B-B)$. This value can be checked by calculating it from other compounds with a boron-boron bond.

The mean bond energies for the covalent bonds of the non-metals are given for each non-metal in the appropriate chapter later in the book. The question now is: To what use can bond energies be put? Firstly they enable $\Delta H_f^\ominus$ to be calculated for a compound when this cannot be measured directly; for instance, with diphosphine, P_2H_4, which has the same structure as B_2Cl_4, with a P–P bond. Using the bond energy data of Table 15, $\Delta H_{atom}^\ominus$ can be calculated:

$$\Delta H_{atom}^\ominus = E(P-P) + 4E(P-H) = 209 + 4 \times 322 = 1497 \text{ kJ mol}^{-1}$$

and this can be used to calculate $\Delta H_f^\ominus(P_2H_4)$ by using equation (1):

$$\Delta H_{atom}^\ominus(P_2H_4) = 2\Delta H_f^\ominus(P) + 4\Delta H_f^\ominus(H) - \Delta H_f^\ominus(P_2H_4)$$
$$\text{therefore } 1497 = 2 \times 315 + 4 \times 218 - \Delta H_f^\ominus(P_2H_4)$$
$$\text{therefore } \Delta H_f^\ominus(P_2H_4) = 5 \text{ kJ mol}^{-1}$$

This value fits well with the observation that diphosphine is an unstable compound, decomposing spontaneously.

Secondly bond energies enable predictions to be made about the likelihood of chemical reactions taking place. Consider the general reaction:

$$AB_n + nC = AC_n + nB$$

which we would expect to occur if $E(A-B)$ were smaller than $E(A-C)$, that is if we were replacing a weaker by a stronger bond, and assuming that all other energy changes are insignificant. Thus to convert PCl_3 to PF_3 should be easy since $E(P-Cl)$ is 319 and $E(P-F)$ is 490 kJ mol^{-1}. In this case a variety of reagents will convert PCl_3 to PF_3. However, care must always be taken when making judgements based solely on bond energy values because other energy changes in a reaction may be more important.

Since $E(N-H)$ is 390 kJ mol^{-1} and $E(N-F)$ is 272 kJ mol^{-1}, one would expect the conversion of NH_3 to NF_3 to be difficult, yet it is quite easily done at room temperature by mixing fluorine and ammonia:

$$2NH_3 + 3F_2 = 2NF_3 + 3HF$$

and the reason lies in the bond dissociation energies of F_2 (153 kJ mol^{-1}) and HF (566 kJ mol^{-1}), and the difference between these more than compensates for the adverse bond energy difference of NH_3 and NF_3.

So far we have seen that the multiplicity of a bond is reflected in its length and its strength and we can make the generalization that single bonds are longer and weaker than double bonds, which in turn are longer and weaker than triple bonds. We now turn to the third property of a bond we wish to know about and that is the position of the bonding electron-pairs, which leads to the concept of electronegativity.

Electronegativity (χ)

It has been known for a long time that certain atoms attract electrons more strongly than others. When such atoms are part of a molecule they will have the effect of increasing the electron density in their vicinity at the expense of other parts of the molecule. These atoms are said to have a high electronegativity. *Electronegativity, χ, is defined as the ability of an atom in a molecule to attract electrons to itself.* The alkali metals have very low electronegativities and are often called electropositive elements for this reason. The halogens have high electronegativities and fluorine is the most electronegative element of all.

There is no way of measuring electronegativity directly, and so it must be approached through properties which are closely linked to it. One such property is the dipole moment, p_e, which is a measure of the distribution of electrons in a molecule. Strictly speaking a dipole moment arises when the centre of positive charge in the molecule due to the protons of the nuclei does not coincide with the centre of negative charge due to the electrons. The dipole moments of many molecules are known and in simple diatomic molecules like HF, HCl, etc., they are directly related to the electronegativity difference between the two atoms involved. But even with diatomic molecules there may be other factors which blur the relationship and we find that carbon monoxide has a low dipole moment even though there is quite a large difference between the χ values of oxygen and carbon. Although dipole moments cannot be measured directly they can be calculated from dielectric constants.*

In molecules with more than two atoms it is still possible to relate p_e to χ, but in many cases this cannot be done. In some molecules such as SiF_4 the dipole moment is zero because of the symmetrical structure of the molecule, although for each individual Si–F bond we know the electrons are closer to the fluorine than the silicon atom. Again, partly for this reason, it is not feasible to derive a set of electronegativity values from dipole moment data. Nevertheless the dipole moments of molecules do tell us something about the overall distribution of electrons in molecules, if not in individual bonds. This information can be very useful. For instance the 'attractiveness' of polar molecules explains a lot of the behaviour of solvents, especially in their ability to dissolve ions. (Why?)

Although dipole moments cannot be used as a basis of electronegativity there are other properties of molecules which reflect χ and so can be used as the foundation for a table of values. Such a table, in which each element is given a numerical value of χ, is obviously desirable. To know that $\chi(F)$ is greater than $\chi(Cl)$ is useful, but more useful still is the knowledge of how much greater $\chi(F)$ is than $\chi(Cl)$. Linus Pauling was the first to draw up such a table and he based his values on bond energies. He noticed

* Dielectric constants are measured by comparing the charge on a capacitor with nothing between the plates and with the material to be measured between the plates. A simple way of demonstrating the effect of an electric field on a liquid is to fill a burette with a solvent such as acetone, open the tap, and bring a piece of charged material (such as a piece of polythene or perspex rubbed against clothing) up to the jet. Solvents with high dielectric constants and high dipole moments will be deflected and these are called polar solvents. Solvents with low dielectric constants and low dipole moments will be relatively unaffected; these are often called non-polar solvents.

that for a covalent bond A—B the bond energy $E(A-B)$ was always greater than the average bond energies of the bonds A—A and B—B, that is:

$$E(A-B) > \tfrac{1}{2}[E(A-A) + E(B-B)]$$

The difference between the observed value $E(A-B)$ and the average $\tfrac{1}{2}[E(A-A) + E(B-B)]$ he called Δ and said that this extra energy of the bond A—B came about because the bond was not purely covalent but was partially ionic in nature. The ionic contribution made the bond stronger than expected. In the ionic bond both of the bonding electrons are associated with one atom:

$$A-B \leftrightarrow \overset{+\,-}{A\,B} \leftrightarrow \overset{-\,+}{A\,B}$$
$$\text{covalent} \quad \text{ionic forms}$$

Both ionic forms are not equally important. If A is more electronegative than B then it will attract the electron-pair more strongly and the $\overset{-+}{AB}$ ionic form will predominate; the $\overset{+-}{AB}$ form can be ignored. The greater the electronegativity difference, $\chi(A) - \chi(B)$, the greater the ionic contribution to the bonding will be and so the greater will be Δ.

$$[\chi(A) - \chi(B)] \propto \Delta$$

From Δ Pauling was able to obtain not electronegativities themselves but electronegativity differences. To give each atom a χ value he chose $\chi(H)$ to be 2·1. This gave him a set of values ranging from $\chi(Cs) = 0\cdot7$, the least electronegative element, to $\chi(F) = 4\cdot0$, the most electronegative element. The χ values for all the main group elements are given in Table 7.

There are other ways of determining electronegativities. A method developed by Mulliken uses ionization potentials and electron affinities.* (An experiment for measuring the ionization potential of xenon is given in Chapter 10.) Another method is that of Allred and Rochow, and this is based on the strength of the nuclear charge felt by electrons at the covalent radius boundary. The table of values obtained by the Mulliken and Allred-Rochow methods are very similar to those reported in Table 7, which incorporates some of these values for those elements to which Pauling's method is not applicable. The table shows that χ increases from left to right of the periodic table and generally decreases down a particular group. This is not true of gallium in M3 and germanium in M4, because with these elements there are underlying filled d shells of electrons which are not as effective in screening the nuclear charge as other shell. Thus the electrons are under the influence of a larger nuclear charge than expected and they thus have larger electronegativities than expected.

* The ionization potential of a neutral atom is the energy required to remove an electron completely from it. The electron affinity of a neutral atom is the energy *given out* when an electron is added to it. Although ionization potentials can be accurately measured and are known for almost all the elements, electron affinities are difficult to obtain and are known accurately for only a few elements.

Knowing the χ values of the elements involved in a bond gives an indication of the distribution of electrons in the bond. This in turn enables predictions about the behaviour of the bond in chemical reactions to be made. Some reagents are known to attack that part of a molecule where electron density is highest—these are called

Table 7 Pauling electronegativities of the main group elements

						H	He
						2·1	—
Li	Be	B	C	N	O	F	Ne
1·0	1·5	2·0	2·5	3·0	3·5	4·0	—
Na	Mg	Al	Si	P	S	Cl	Ar
0·9	1·2	1·5	1·8	2·1	2·5	3·0	—
K	Ca	Ga	Ge	As	Se	Br	Kr
0·8	1·0	1·8*	2·0*	2·1	2·4	2·8	3·0*
Rb	Sr	In	Sn	Sb	Te	I	Xe
0·8	1·0	1·5*	1·7*	1·8	2·1	2·4	2·8*
Cs	Ba	Tl	Pb	Bi			
0·7	0·9	1·4*	1·6*	1·7			

* These values were obtained by the Allred-Rochow method.

electrophilic reagents. Reactants carrying a positive charge are usually strong electrophilic reagents, e.g. NO_2^+. Conversely some reagents prefer to attack centres of the molecule where the electron density is lower and these are called nucleophilic reagents. Examples of such reagents are negatively charged species such as OH^-, Cl^-, or reactants which have an active non-bonding electron-pair in the molecule such as NH_3.

3 The structure of molecules

In the previous chapter we discussed the covalent bond. No account was taken of the bond's environment, that is of other atoms and bonds within the same molecule, except in the rather specialized case of delocalization. In this chapter we shall look at the structure of a molecule as a whole. In speaking of the structure, or as it is often called the stereochemistry of a molecule, we shall need to use the terms *coordination number* and *configuration*.

Coordination number (c.n.)

An atom in a molecule, or molecular ion, can be covalently bonded to as many as eight other atoms. This number of atoms is called the coordination number of the central atom. Eightfold c.n. is extremely rare and sevenfold even rarer, but c.n.'s of one to six are very common. In the ammonia molecule, NH_3, the nitrogen atom has c.n. 3 and this is the most common c.n. of nitrogen. In hydrogen cyanide, HCN, it has c.n. 1, in imine derivatives, RNH, it has c.n. 2, and in the ammonium ion, NH_4^+,

Table 8 Coordination numbers of the non-metals

Row	Maximum c.n.	Non-metals and their common c.n.'s					
1	2	H (1, 2)					
2	4	B (3, 4)	C (3, 4)	N (3, 4)	O (1, 2)	F (1)	
3	6		Si (4)	P (3, 4)	S (2, 3, 4)	Cl (1)	
4	6				Se (3, 4)	Br (1)	Kr (2)
5	8					I (1, 3, 4)	Xe (4, 6)

it has c.n. 4. Nitrogen never shows a higher c.n. than four. Iodine on the other hand can exhibit a range of c.n.'s from one to seven as in the following series of compounds; I_2 c.n. 1, ICl_2^- c.n. 2, ICl_3 c.n. 3, IF_4^+ c.n. 4, IF_5 c.n. 5, IF_6^- c.n. 6, and IF_7 c.n. 7.

The coordination number of an atom in a molecule is not to be confused with its valency. There is a connection between the two but it is best to treat c.n. simply as its name implies and that is as a number which tells us something about the structure of the molecule. Although the c.n. of an atom can vary, not all have the versatility of iodine and for each element there is a limit to the maximum c.n. This limit depends upon the row of the periodic table to which the atom belongs. Table 8 gives the maximum c.n.'s of the non-metals together with the most common c.n.'s displayed by them in their compounds.

Configuration

Simply knowing the coordination number of all the atoms in a molecule is not sufficient to give a complete picture of the molecule. The configuration of the molecule must also be known, in other words how the atoms are arranged in space with respect to one another. For example, the compound BF_3 in which the c.n. of boron is 3 could have two possible configurations shown in Fig. 5. In the first of these (I), all the atoms are in the same plane and the configuration is called trigonal *planar*. In the second (II), the atoms form a pyramid with boron at the apex and the three fluorines forming a triangular base. This form is known as *trigonal pyramidal* or simply pyramidal. The word trigonal signifies that the base of the pyramid is a triangle, and there are structures where the base can be a square or even a five-sided figure (called a pentagonal pyramid). Of the two structures shown in Fig. 5, boron trifluoride has the planar

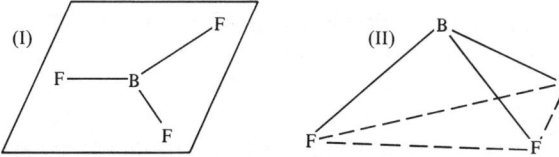

Fig. 5. The possible configurations of BF_3

one (I). Structure (II) is also quite common and is found in nitrogen and phosphorus compounds in which these have c.n. 3, e.g. NH_3 and PCl_3.

Knowing that there are many structures which molecules can have, the natural question to ask is what causes a molecule to adopt a particular configuration. There are two theories which attempt to explain the stereochemistry of molecules. One is called the valence bond theory and is based on the formation of hybrid orbitals from atomic orbitals, and the direction of these hybridized orbitals in space determines the configuration of the molecule. The valence bond theory seeks not only to explain the stereochemistry of molecules but also to explain the nature of covalent bonding in general. Though popular in the years 1930–1960 this theory has been widely attacked in the past decade and is rapidly falling out of favour.

Partly to fill the gap left by the valence bond theory, a much simpler theory based on electron-pair repulsions has been developed. This is the Sidgwick-Powell theory which was first published in 1940, and although it is a simple theory it gives an adequate explanation of the structures of all known non-metal compounds. The theory considers the electrons of the valence shell of the central atom, and these are either bonding or non-bonding electron-pairs. Bonding electron-pairs we have already met; non-bonding pairs are also in the outer valence shell but, as their name implies, they are not involved in the covalent bonding between two atoms. They are sometimes called lone pairs and they have an important role in much of the chemistry of the non-metals.

The bonding and non-bonding electron-pairs arrange themselves so as to reduce coulombic repulsions to a minimum. Bonding pairs are further from the nucleus of the central atom than non-bonding pairs, which is to be expected since bonding pairs are under the influence of a second nucleus which is also exerting an attraction upon them. Non-bonding pairs are more diffuse in space than bonding pairs and this also has an effect on the repulsions between electron pairs. In order of magnitude these repulsions are:

non-bonding ↔ non-bonding > non-bonding ↔ bonding > bonding ↔ bonding

and their effect can be seen when the bond angles of methane, ammonia, and water are compared (Fig 6). The reader is invited to draw the structures and make predictions about the bond angles of other molecules with two, three, four, five, six, seven, and eight electron-pairs before continuing with this chapter.

A molecule with only two bonding pairs will have these arranged at opposite sides of the central atom and the molecule will be linear, as in beryllium dichloride:

$$Cl-Be-Cl$$

Three bonding electron-pairs will arrange themselves in a plane at angles of 120° to one another, as in boron trifluoride (Fig. 5(I)). In neither of these cases do we encounter non-bonding pairs and it is only when we come to four or more electron-pairs surrounding the central atom that we meet non-bonding pairs. The structures found with four, five, and six electron-pairs (when they may all be bonding pairs, or one or two may be non-bonding pairs) are shown in Fig. 6, which gives actual compounds in which these structures are found. In some cases, notably sulphur tetrafluoride, SF_4, chlorine trifluoride, ClF_3, and xenon tetrafluoride, XeF_4, there are other arrangements possible for the lone pairs, but these all have more coulombic repulsion associated with them than the structures shown in Fig. 6. For example, the other alternative arrangement for SF_4 would have the non-bonding pair in the vertical position and at right-angles to three fluorine atoms. This structure would have a larger repulsion factor than that illustrated in which the non-bonding pair is at right-angles to two fluorine atoms. The reader should draw the other possible configurations for ClF_3 and XeF_4 and prove to his own satisfaction that these would be less stable than the illustrated structures.

With seven electron-pairs, as in IF_7, and eight electron-pairs, as in XeF_8^{2-}, the configurations are pentagonal bipyramidal and cubic as shown in Fig. 7.

When the bonding involves π-electron-pairs, the Sidgwick-Powell theory needs slight modification. However, as a general rule it is only the electron-pairs of the σ-bonds which determine the structure, so that carbon dioxide, CO_2, is linear like $BeCl_2$, while the carbonate ion, CO_3^{2-}, is planar with an OCO bond angle of 120° like

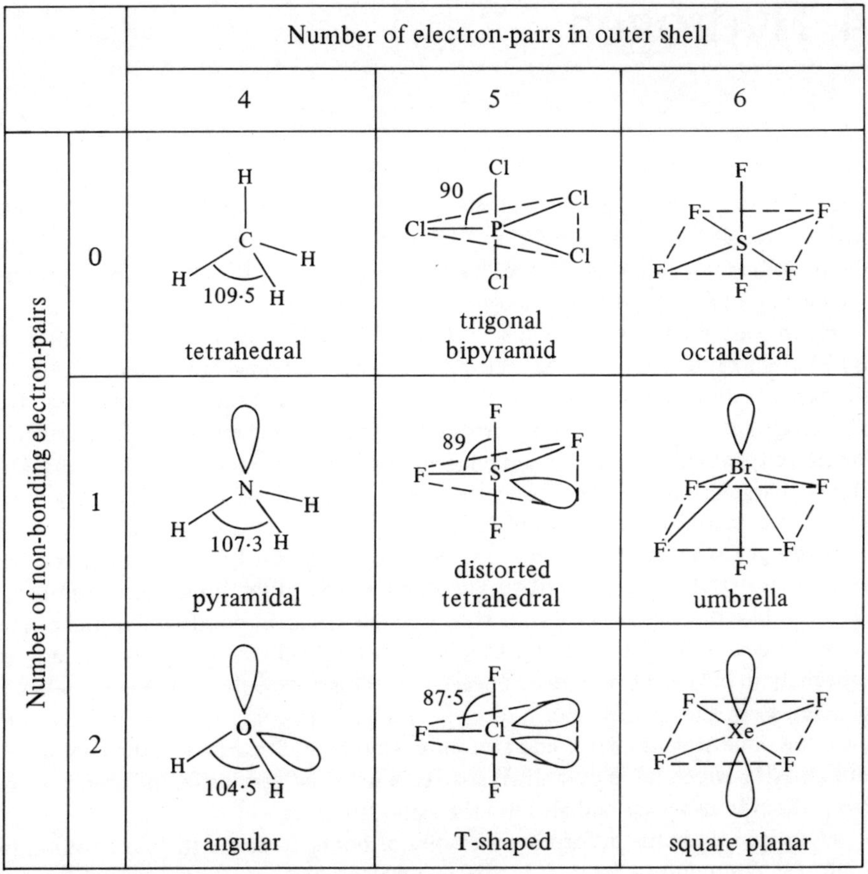

	Number of electron-pairs in outer shell		
	4	5	6
0	tetrahedral (CH₄, 109·5°)	trigonal bipyramid (PCl₅, 90°)	octahedral (SF₆)
1	pyramidal (NH₃, 107·3°)	distorted tetrahedral (SF₄, 89°)	umbrella (BrF₅)
2	angular (H₂O, 104·5°)	T-shaped (ClF₃, 87·5°)	square planar (XeF₄)

Fig. 6. The structures of non-metal compounds

BF₃. In these species there is only one bonding pair of electrons per bond to be considered so in effect CO₂ has only two σ-bonding electron-pairs and CO₃²⁻ has only three σ-bonding electron-pairs.

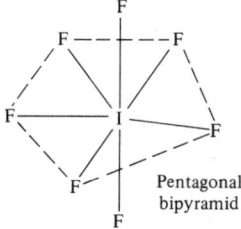

Fig. 7. The structure of IF_7 and XeF_8^{2-}

4 Hydrogen

The chemistry of hydrogen bears little resemblence to that of any other element, with the result that it cannot fittingly be placed at the head of any of the groups of the periodic table, although it does have some properties in common with groups M1, the alkali metals, and M7, the halogens.

Hydrogen, like the alkali metals, has one s electron in its valence shell and there is likewise a tendency to lose this electron to give a monovalent cation, H^+. The similarity is more apparent than real, however, because the loss of an electron from a hydrogen atom produces a bare proton which is not a chemical entity as such. When we speak of H^+ in acidic aqueous solutions we are really referring to H_3O^+ or $H_9O_4^+$. Large differences are shown between hydrogen and the alkali metals in their m.p.'s, b.p.'s, ionization potentials, and many other physical properties.

Hydrogen is generally placed at the head of group M7. Like the halogens it is one electron short of a rare gas configuration and will form the simple monatomic anion, H^-. Also like the halogens it forms strong, single covalent bonds and in the elemental state exists as a diatomic gas, H_2. Even so it differs from the halogens in many ways, especially in having a lower electronegativity. This means that there are fundamental differences in the chemistry of hydrogen and halogen compounds even when these have the same stoichiometry and structure. Compare, for instance, the chemistry of the gas NH_3 which has a pyramidal structure with the very different chemistry of NF_3 which is also a gas and also has the same structure.

Hydrogen is unique in forming six types of bond. Basically there are four kinds of hydrogen compound – ionic, interstitial, covalent, and acidic. In addition there are two special types of bonding called hydrogen bonding or *H-bonding* and hydrogen bridging or *H-bridging*. The former is encountered in compounds in which the hydrogen is covalently bonded to the elements of high electronegativity such as fluorine, oxygen, and nitrogen. The latter is found in the hydrogen derivatives of the 'electron deficient' elements beryllium and boron.

Ionic hydrides

Only the more electropositive metals of groups M1 and M2 form ionic hydrides which contain the hydride ion H^-. This has only two electrons and an ionic radius of 140 pm, which is larger than that of the fluoride ion (133 pm) with its ten electrons. The electron density is accordingly very low and the ion is easily polarized and its electron cloud deformed in the presence of small cations like those of the electropositive metals. The result is partial overlap between the cation and anion and this gives a certain covalent character to the bonding.

The ionic hydrides are formed by the action of hydrogen on the molten metal at temperatures above 500 K. They are white solids which are thermally stable, but they react rapidly with water and hydrogen gas is released:

$$MH + H_2O = MOH + H_2$$

For this reason they are unstable in air, although sodium and calcium hydrides which are commercially available have a long shelf-life. These hydrides are grey due to the presence of unreacted metal. They are commonly used for drying organic solvents.

Interstitial hydrides

This name is given to the hydrides of the transition metals (d-block elements). The bonding in these compounds is not clearly understood but appears to consist of

Table 9   Bond energies, E, and bond lengths, r, of hydrogen bonds

Bond	E(kJ mol^{-1})	r/(pm)	Bond	E/(kJ mol^{-1})	r/(pm)
H–H*	436	75	H–O	467	96
H–B	381	119	H–S	347	133
H–C	416	108	H–Se	276	147
H–Si	323	146	H–F*	566	93
H–N	390	101	H–Cl*	431	128
H–P	322	142	H–Br*	366	142
			H–I*	299	162

* bond dissociation energy, D

hydrogen atoms in the holes or interstices in the lattice of metal atoms (hence the name interstitial). Since hydrogen atoms are small they can occupy these holes without causing much change in the structure of the lattice. Thus the bulk properties of the metal such as hardness remain unchanged. Not all holes in the lattice of a transition metal hydride contain a hydrogen atom, so that the compounds are not stoichiometric and the metal:hydrogen ratio can vary within certain limits. Formulae are normally written in the manner TiH$_{1.7}$ and ZrH$_{1.9}$, for example. Copper hydride has the stoichiometric formula CuH and this compound resembles both the interstitial and ionic hydrides. (Why is this?)

Covalent hydrides

Principally these are the hydrides of the non-metals, but the metals of groups M3, M4, and M5 also form covalent hydrides. The bond energies and bond lengths of the non-metal hydrides are given in Table 9 which shows that all these bonds are quite strong. The bond of hydrogen fluoride is one of the strongest single bonds known.

The electronegativity of hydrogen is 2·1 and only boron and silicon of the non-metals have lower values. This means that the polarity of most hydrogen bonds have the hydrogen atom as the positive end of the bond dipole. The B—H and Si—H bonds on the other hand have the hydrogen atom as the negative end of the dipole. This is reflected in the reaction of these bonds with water when hydrogen is produced as with the ionic hydrides.

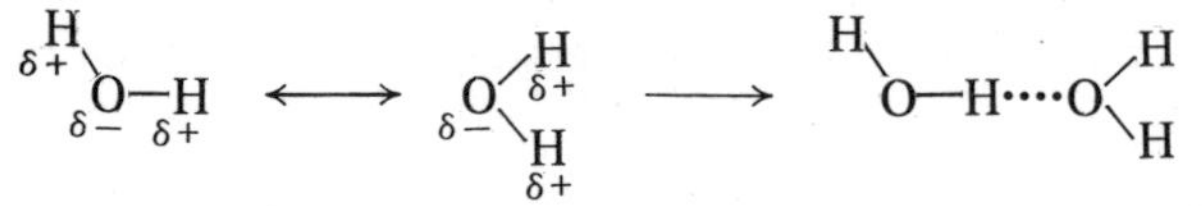

Fig. 8. H-bonding in water

The chemistry of the non-metal hydrides comprises a substantial portion of the chemistry of each non-metal, and so are discussed in later chapters. However there are two features of certain covalent hydrides which are best introduced here. These are H-bonding and H-bridging.

H-bonding is found with compounds in which hydrogen is covalently bonded to a strongly electronegative element such as nitrogen, oxygen, or fluorine, so that there is a strong dipole associated with the bond. The hydrogen, with its net positive charge, becomes attracted to another electronegative atom in its vicinity and especially towards the non-bonding electron-pairs which such atoms invariably possess. In the case of water this is illustrated in Fig. 8, where the H-bond is shown by a dotted line. Such bonds are weak with energies of about 20 kJ mol^{-1}, and are longer than normal covalent bonds. Nevertheless they have considerable influence on the physical properties of the compounds in which they occur, as Fig. 9 shows. This is a plot of the b.p.'s of the hydrides of the non-metal hydrides against their molecular weights. As a rule the b.p. of a series of similar compounds increases as the molecular weight increases. This is illustrated by the b.p.'s of the rare gases and the hydrides of group M4 elements which do not form H-bonds. The other groups show the same trend except for the hydrides of the strongly electronegative elements, i.e. NH_3, H_2O, HF, and to a lesser extent HCl. These have higher b.p.'s than would be predicted from other members of the group. The higher b.p.'s stem directly from H-bonding in the liquid phase and this is also responsible for the higher heats of evaporation of these compounds.

A simple experiment can be devised for showing the strength of an H-bond. In this two liquids which react to form an H-bond are mixed and the rise in temperature is recorded. Trichloromethane is capable of H-bonding even though the hydrogen is attached to a carbon atom, because the three chlorine atoms of Cl_3CH increase the effective electronegativity of the carbon atom. An ideal pair of liquids for studying H-bond formation are thrichloromethane and acetone. Full details can be found in *The Student's Book I. Chemistry. Nuffield Advanced Science.*

H-bridging is found in compounds which are 'electron deficient' such as beryllium hydride and the boron hydrides. In these compounds the bond is not electrostatic in nature as is the H-bond, but involves the overlap of the hydrogen orbital with the orbitals of *two* boron atoms. Into the new orbital so-formed go two electrons which serve to bind the three atoms. For this reason hydrogen is said to be bridging the two boron atoms. Diborane, which is the best-understood example, is discussed in more detail in the next chapter. H-bridge bonds have bond energies of the same magnitude as normal electron-pair covalent bonds.

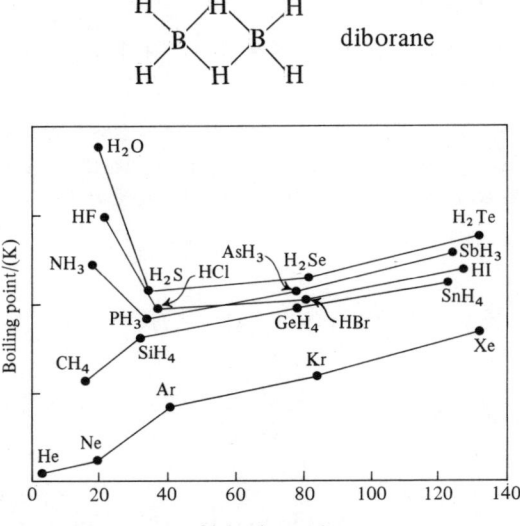

Fig. 9. Boiling points of the non-metal hydrides and rare gases

Acidic hydrides

These hydrides are often called protonic acids, or simply acids. In this type of hydrogen compound the proton is capable of detaching itself from the parent molecule, which is called the acid, and attaching itself to another molecule called the base. The base then in effect becomes an acid and is given the name *conjugate acid*. The system can be thought of as an equilibrium:

$$HA + B \rightleftharpoons A^- + HB^+$$
$$\text{acid} \quad \text{base} \quad \text{conjugate base} \quad \text{conjugate acid}$$

which can be divided into two half-reactions, one for the acid and conjugate base and the other for the base and conjugate acid:

$$HA \rightleftharpoons A^- + H^+$$
$$H^+ + B \rightleftharpoons HB^+$$

H^+ is in fact a bare proton which is not a chemically meaningful entity and these half-reaction equations do not represent actual equilibria. The sum of these two equations is the acid-base equilibrium.

The most acidic hydrogen atoms are those attached to halogen or oxygen atoms, but under the right basic conditions even the most unlikely hydrogen can behave in an acidic manner. Acetylene, for example, in liquid ammonia has an acidic hydrogen. Conversely, molecules which are normally thought of as acids can be induced to act as bases and accept a proton. For instance phosphoric acid in pure sulphuric acid as the solvent behaves this way:

$$H_3PO_4 + H_2SO_4 \rightleftharpoons H_4PO_4^+ + HSO_4^-$$
$$\text{base} \quad\quad \text{acid}$$

When water is used as a solvent the proton attaches itself to a water molecule which acts as the base to give the oxonium ion, H_3O^+. In the bulk liquid this ion forms three H-bonds to three other water molecules to give the species $H_9O_4^+$.

H-bonding plays a very important part in the chemistry of protonic acids in water. The proton is apparently very mobile in water and this is due to H-bonding, which facilitates a bonding rearrangement between adjacent water molecules with the result that the proton appears to move:

$$\overset{+}{H-O}-H\cdots O-H\cdots O-H\cdots O-H \longrightarrow H-O\cdots H-O\cdots H-O\cdots H-\overset{+}{O}-H$$
$$\quad | \quad\quad | \quad\quad | \quad\quad | \quad\quad\quad\quad\quad | \quad\quad | \quad\quad | \quad\quad |$$
$$\quad H \quad\quad H \quad\quad H \quad\quad H \quad\quad\quad\quad\quad H \quad\quad H \quad\quad H \quad\quad H$$

Hydrogen isotopes

Hydrogen has three isotopes, ^1H, ^2H, and ^3H, having one proton and none, one or two neutrons respectively in the nucleus. These isotopes are given the chemical symbols H (hydrogen), D (deuterium), and T (tritium).

Deuterium oxide, D_2O, is used in chemistry to study the exchange reactions which occur between hydrogen attached to a compound and the solvent water in which it is dissolved. For example, when methylammonium chloride, CH_3NH_3Cl, is dissolved in D_2O and then reclaimed, it is found to have exchanged only the three hydrogen atoms attached to the nitrogen atom to give CH_3ND_3Cl. The hydrogen atoms of the methyl groups remain. (The reader should be able to suggest the reason for this behaviour).

The molecular weight of H_2O is 18·01 and that of D_2O is 20·01. This 10% increase

Table 10 The physical properties of H_2O and D_2O

	H_2O	D_2O
M.p./(K)	273	277
B.p./(K)	373	374
Specific gravity at 273 K	0·9982	1·1059
Temperature of maximum density/(K)	277	285
$\Delta H°_{fusion}$ at m.p./(kJ mol^{-1})	6·00	6·37
$\Delta H°_{vaporization}$ at b.p./(kJ mol^{-1})	40·57	41·68

is reflected in many of the physical properties as Table 10 shows. In most cases, substituting deuterium for hydrogen has only a small effect on physical properties of molecules because the weight increase is proportionately very small.

The rarer tritium isotope is radioactive with a half-life of 12·5 years decaying by β-emission. It too finds use in modern chemical research.

5 Boron

Boron is the only non-metal of group M3 and it is the only non-metal that is 'electron deficient'. This means that even when all its valence electrons are involved in covalent electron-pair bonding the boron atom still has not achieved a rare gas configuration. The electronic configuration of boron is $1s^2 2s^2 2p^1$ which suggests a valency of one in which only the $2p^1$ electron is involved. At the bottom of group M3 is thallium and this does in fact behave in this way. However, boron displays a valency of three in all its compounds and it achieves this by promoting one of the $2s^2$ electrons to a $2p$ orbital:

$$1s^2 2s^2 2p^1 \rightarrow 1s^2 2s^1 2p^2$$

There are consequently three unpaired electrons in the outer shell capable of forming three covalent bonds. It requires a further two electrons to complete the rare gas configuration and these it is willing to accept from another molecule or ion. In so-doing boron compounds are behaving as *Lewis acids*.

LEWIS ACIDS AND BASES A Lewis acid is a compound which will accept a non-bonding electron-pair and a Lewis base is a compound which will donate a non-bonding electron-pair. When a Lewis acid reacts with a Lewis base a product is formed which is called an *adduct*. The new bond formed between the Lewis acid and Lewis base is a covalent bond which has all the characteristics of a normal electron-pair bond, except that both electrons were supplied by one of the atoms. Trivalent boron compounds are generally good Lewis acids, while trivalent nitrogen compounds make good Lewis bases. A typical reaction of this kind is that of BF_3 and $N(CH_3)_3$:

$$BF_3 + N(CH_3)_3 = F_3B \cdot N(CH_3)_3$$

An alternative way of denoting the adduct is to use a symbol which suggests the donating of electrons, such as $F_3B \leftarrow N(CH_3)_3$ or $F_3\overset{-}{B}-\overset{+}{N}(CH_3)_3$.

Many aluminium derivatives are good Lewis acids and aluminium chloride acts as a Lewis acid through the aluminium atoms and as a Lewis base through the non-bonding electron-pairs on the chlorine atoms. The result is a dimer:

$$\begin{array}{ccccc} Cl & & Cl & & Cl \\ & \diagdown & | & \diagdown & \\ & & Al & & Al \\ & \diagup & | & \diagup & \\ Cl & & Cl & & Cl \end{array}$$

The tetrafluoroborate anion, BF_4^-, can also be seen as an adduct of the Lewis acid BF_3 and the Lewis base F^-:

$$BF_3 + F^- = BF_4^-$$

Trivalent boron derivatives have a planar arrangement since there are three electron-pairs surrounding the boron atom, see Fig. 10. Boron in Lewis adducts has four electron-pairs and so adopts a tetrahedral configuration. The reader is invited to deduce the structure of Lewis acid-base dimers like Al_2Cl_6.

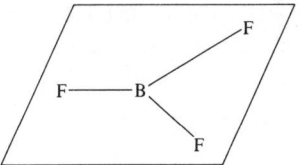
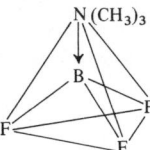

Fig. 10. The structures of boron compounds

Compounds of boron

Table 11 lists the bond energies and bond lengths of the more common covalent boron bonds. From bond energy data alone one might expect BF_3 to be very stable. However, because of the Lewis acidity of this molecule it is very reactive. Bond energies cannot be used to predict the reactivity of a molecule. They allow predictions about the thermodynamic stability of a molecule to be made but say nothing about its kinetic inertness. In the case of BF_3, and many other trivalent boron compounds, reaction with any molecule having an active non-bonding valence pair of electrons will be facilitated by the primary formation of a Lewis adduct.

Table 11 Bond energies, E, and bond lengths, r, of boron bonds

Bond	E/(kJ mol^{-1})	r/(pm)	Bond	E/(kJ mol^{-1})	r/(pm)
B—H*	381	119	B—F	644	129
B—O	523	136	B—Cl	444	174
B—B	335	175	B—Br	368	187
B—C	372	156	B—I	272	203

* normal electron-pair bond

BORON–HYDROGEN In the early years of this century it was observed that the addition of a metal boride (formed by heating the metal with boron) to an acid produced foul-smelling gases which were spontaneously inflammable in air and burned with a green flame. The study of these compounds by Stock had two important consequences for chemistry. First he separated and characterized six boron hydrides; secondly, and in the long run just as important, he developed a method of handling these compounds in evacuated apparatus. His techniques have been used ever since for the investigation of any gaseous or volatile compound which is sensitive to air.

Just over a dozen boron hydrides, or *boranes,* have been isolated so far. The simplest is diborane, B_2H_6, and the most complex is icosaborane, $B_{20}H_{16}$. The monomer, BH_3, itself does not exist, even in equilibrium with B_2H_6 at high temperatures. The chemistry of the boranes is exceedingly involved and this in part stems from the ability of boron to form 'cages' made up of ten to twelve boron atoms. Diborane, however, is relatively simple and is typical of most of the boranes. Nowadays it is prepared by the reaction of BF_3 and sodium borohydride:

$$BF_3 + 3NaBH_4 = 2B_2H_6 + 3NaF$$

Diborane is a gas (b.p. 180 K) and it has the structure shown in Fig. 11, in which four of the hydrogens are bonded to boron in normal single bonds, and two are

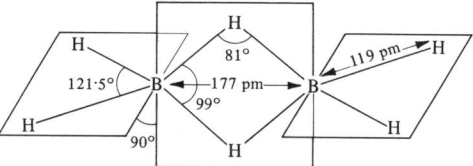

Fig. 11. The structure of diborane

involved in H-bridging. These lie in a plane at right-angles to the normal B–H bonds. The molecule contains twelve valence electrons, i.e. six electron-pairs. Four of these are used by the four normal B–H bonds, leaving only one electron-pair per B–H–B bridge bond. In other words three atoms are held together by only two electrons. Such bonds are often called three centre bonds. One way of describing the bonding involves the overlap of hybridized boron orbitals (sp^2) with the 1s orbital of hydrogen. This produces a bonding orbital which embraces the hydrogen nucleus, and into which goes the electron-pair. The bond energy of the B–H–B bond is 440 kJ mol^{-1} which is slightly stronger than the normal B–H bond. This is demonstrated by the reaction of diborane and bromine:

$$B_2H_6 + Br_2 = B_2H_5Br + HBr$$

in which the hydrogen of a normal B–H bond is replaced by bromine while the two H-bridge bonds remain intact. With chlorine on the other hand the reaction does not stop at this stage but proceeds to boron trichloride, BCl_3.

Another type of reaction of diborane is that with Lewis bases when the product is almost invariably an adduct, e.g.:

$$B_2H_6 + 2N(CH_3)_3 = 2H_3B \cdot N(CH_3)_3$$

These reactions can be imagined as going through an initial step which involves the formation of BH_3, which would be a good Lewis acid, but this step has not been proved.

In a similar sort of reaction with sodium hydride the product is the very useful reducing agent $NaBH_4$:

$$B_2H_6 + 2NaH = 2NaBH_4$$

Sodium borohydride is a milder reducing agent than the other reagent of this kind, lithium aluminium hydride, $LiAlH_4$, but has the added advantage of a long shelf-life. Sodium borohydride is an ionic compound containing the BH_4^- anion, which has the same number of electrons as methane, CH_4 (i.e. they are isoelectronic), and the ammonium ion, NH_4^+, and likewise has the same tetrahedral structure. Sodium borohydride is manufactured by the reaction of NaH and trimethoxyboron, $B(OCH_3)_3$:

$$4NaH + B(OCH_3)_3 = NaBH_4 + 3NaOCH_3$$

At first sight this reaction seems very unfavourable in terms of boron bond energies, but a little thought will indicate where the energy for the reaction is coming from.

BORON—FLUORINE   Although all the boron halides are known the fluoride is particularly interesting. To begin with, the value of E(B—F) is much larger than expected from the bond energy values of the other boron-halogen bonds (Table 11). The reason for this is that the B—F bonds have a certain amount of π-bond character. The π-bond is formed by the overlap of an empty p orbital on boron with a filled p orbital on one of the fluorine atoms. This is in effect a donor π-bond analogous to a Lewis adduct in which the boron atom is acting as an acid and accepting one of the non-bonding electron pairs of a fluorine atom. Since all the fluorine atoms are equivalent all are involved in this extra bonding and it is better to imagine a delocalized π-bond covering the whole molecule. The effect of the π-bonding is threefold; firstly E(B—F) is larger than for just a single bond; secondly the bond length, r, is shorter than expected; and thirdly BF_3 should be a weaker Lewis acid than the other boron

trihalides because it has partially satisfied its electron defficiency by the donor π-bonding. This is indeed so and, of the trihalides, the trifluoride is the weakest Lewis acid.

Boron trifluoride can easily be prepared from boric oxide, B_2O_3, and hydrogen fluoride, HF. The latter is not used as such but can be generated in the reaction flask from calcium fluoride and sulphuric acid. On contact with air, BF_3 is hydrolyzed to give a dense fog of the products of hydrolysis.

With water, BCl_3, BBr_3, and BI_3 are rapidly hydrolyzed to boric acid $B(OH)_3$, as one might expect from a consideration of bond energy data. Boron trifluoride is soluble in water and also gives some $B(OH)_3$, but the major product is tetrafluoroboric acid, HBF_4:

$$4BF_3 + 3H_2O = 3HBF_4 + B(OH)_3$$

In this reaction the total number of B–F bonds remains unaltered. Although the tetrafluoroboric acid is not known in the pure state many of its salts are known.

BORON–OXYGEN   In most boron-oxygen compounds the boron atom has a planar arrangement of three oxygen atoms about it. In the oxide, B_2O_3, each oxygen atom links two boron atoms and the result is a giant molecule with no discrete units. Boric oxide is obtained by dehydrating boric acid:

$$2B(OH)_3 = B_2O_3 + 3H_2O$$

Boric acid has the empirical formula $B(OH)_3$ or H_3BO_3, but in the solid state and in solution it consists of $B(OH)_3$ molecules linked by H-bonds to give much larger units. It is a weak acid in water and it behaves as an acid, not by releasing one of its own hydrogen atoms, but by accepting OH^- from water and displacing the self-ionization equilibrium of water:

$$B(OH)_3 + H_2O \rightleftharpoons B(OH)_4^- + H^+$$

In addition to this primary equilibrium there are secondary equilibria which involve much more complex boron species, again held together by H-bonds.

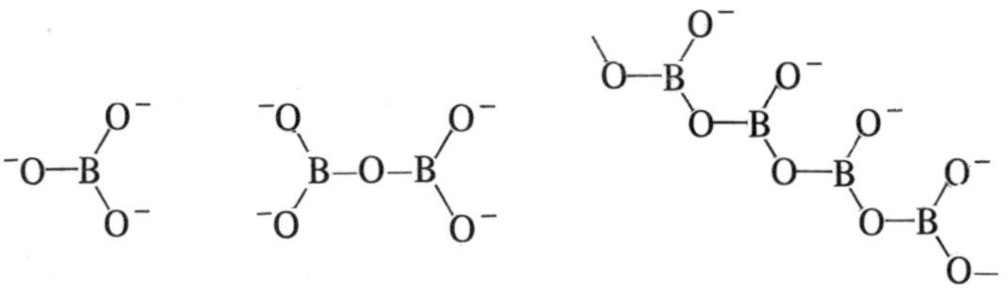

Fig. 12. Structures of the borates

On heating boric oxide and metal oxides, borates are formed. In these compounds the structural unit is BO_3 in which one or two of the oxygen atoms are shared. If none of the oxygen is shared we have the simplest borate with the BO_3^{3-} unit which is found in scandium borate, $ScBO_3$, shown in Fig. 12(I). If one of the oxygen atoms is shared between two boron atoms the result is the $B_2O_5^{4-}$ anion (Fig. 12(II)) and this is found in cobalt borate, $Co_2B_2O_5$. Further sharing of oxygens gives chain-like arrangements (III) and these are present in calcium borate, CaB_2O_4. Even more complex structures are known in which there is a tetrahedral arrangement of oxygen atoms round boron.

6 Carbon and silicon

Of all the elements, carbon and its compounds have been most studied and constitute organic chemistry. Nevertheless there are aspects of the chemistry of carbon which are not dealt with by organic chemistry and these fall within the compass of inorganic chemistry. In particular elemental carbon, the carbon oxides, and such species as the cyanide, cyanate, and thiocyanate anions belong to this group.

For a truly comparative account of carbon and silicon it is necessary to look at many compounds of carbon other than the few just mentioned. The alkanes, alkyl halides, and many others have silicon analogues but often a similarity of structure and stoichiometry is about as far as the comparison goes. Despite the fact that both carbon and silicon are members of group M4 they do not have a lot in common. This dissimilarity between the first and second members of a main group is not confined to group M4. For all the main groups the difference between first and second members is quite pronounced. There are several reasons for this.

Variable valency The second-row elements display only one valency, except under extreme or unusual conditions, whereas non-metals of the third row display two or more valencies. This particularly applies to phosphorus, sulphur, and chlorine and to a lesser extent silicon.

Maximum coordination number The second-row elements have a maximum coordination number of four, the third-row elements six.

Multiple bonds The second-row non-metals can form double and occasionally triple bonds by the use of $2p$ orbitals which overlap to form π-bonds. The third-row elements cannot form such bonds since $3p$ orbitals are unsuitable for this purpose. A much weaker π-bonding involving $3d$ orbitals is known, however, for some of the third-row elements, especially silicon, phosphorus, and sulphur.

Carbon $1s^2 2s^2 2p^2$

Elemental carbon exists in two forms, in other words it displays the property of allotropy. These allotropes are diamond and graphite. The former is so rare on this planet as to be a symbol of wealth, and even though it is now artificially made its manufacture is costly and the quality of the product is visually poor in comparison to that found in nature. In diamond there is a tetrahedral arrangement of carbon as Fig. 13 shows, and all the bond lengths are the same.

The other form of carbon is graphite in which the carbon atoms are in layers of linked six-membered rings. Here the bonding is like that of benzene and all the bond lengths in the planes are the same. The carbon-carbon distance between the planes is much larger, as Fig. 13 shows. Somewhat surprisingly, graphite is the more thermally stable allotrope of carbon even though it is more reactive chemically. It is known for its lubricating power which comes from its planar structure and the presence of small molecules such as oxygen between the layers. These small molecules weaken the bonding between the planes and allow them to glide over one another (cf. the action of ball bearings). If a sample of graphite is put under vacuum for several hours so that these small molecules are removed then the lubricating power of the graphite is considerably reduced.

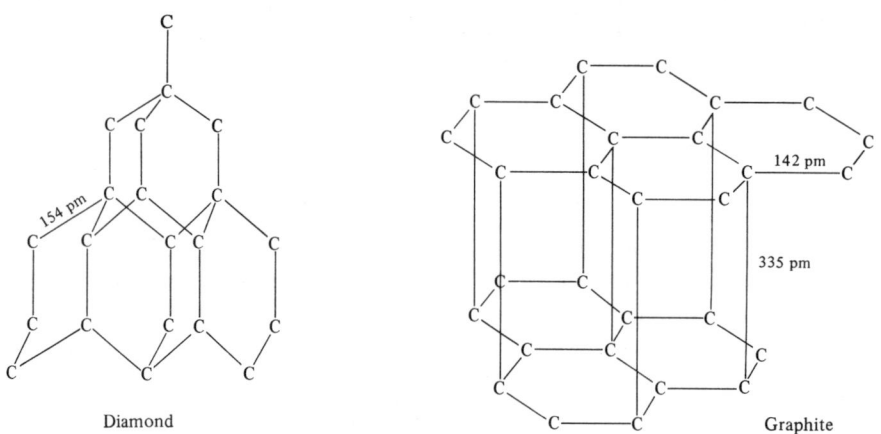

Fig. 13. The structures of diamond and graphite

The planar structure of graphite shows itself in the flat plate-like nature of the crystals which can be seen with the aid of a magnifying glass, or better still a microscope. By using an electron diffraction tube the structure of graphite can be determined exactly and the carbon-carbon distances in the crystal measured. The tube, TEL 555, made by Teltron Ltd has a collimated beam of electrons striking a thin graphite target which causes diffraction and produces two rings on a luminescent screen. The atoms of carbon are acting as a grating and the diameters of the two rings are directly related to the carbon-carbon bond length in graphite. The relationship between the wavelength of the cathode rays, λ, and the radius of the ring, R, gives the interatomic distance, d:

$$\lambda = d \sin \theta = d \frac{R}{L}$$

θ is the angle of diffraction and, if this is small, $\sin \theta$ is equivalent to R/L, where L is the distance from the graphite to the screen of the diffraction tube, as shown in Fig. 14.

Carbon has the most versatile bonding of any element and in addition forms strong covalent bonds with itself, as Table 12 shows. This leads to the formation of stable chains and rings and gives the chemistry of carbon its diversity. The formation of chains composed of the same type of atom is called *catenation* and is not solely confined to carbon, although carbon chains are the most stable. Silicon and sulphur also form chains.

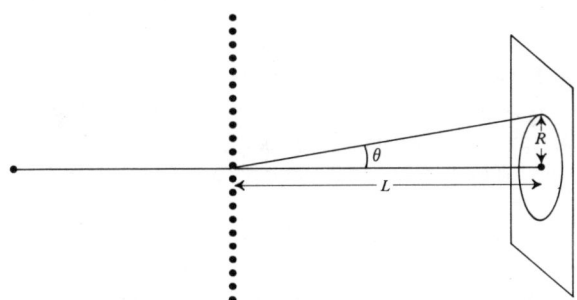

Fig. 14. Electron diffraction tube

CARBON–OXYGEN   Carbon monoxide, CO, is isoelectronic with N_2 but is much more reactive. Table 12 gives its dissociation energy and this shows it to be one of the strongest covalent bonds known. Since there is an electronegative difference of 1·0 between carbon and oxygen a large dipole moment for CO would be expected,

Table 12   Bond energies, E, and bond lengths, r, of carbon bonds

Bond	$E/(kJ\ mol^{-1})$	$r/(pm)$	Bond	$E/(kJ\ mol^{-1})$	$r/(pm)$
C–H	416	108	C–O	358	143
C–F	485	132	C=O†	745	122
C–Cl	327	176	C≡O*	1070	113
C–Br	285	194	C–C	346	154
C–I	213	214	C=C	611	134
			C≡C	837	120

* bond dissociation energy, D
† ketones

but in fact the dipole moment is very low. This is a clue to the bonding, which the D and r values suggest is a triple bond. For oxygen to participate in such a bond it must use one of its non-bonding pairs in the formation of a donor π-bond by the overlap of an empty $2p$ orbital on carbon with a filled $2p$ orbital on oxygen (cf. BF_3). The result of this is to reduce the dipole moment. Both the carbon and oxygen atoms have a non-bonding pair of electrons:

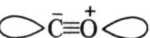

The presence of the non-bonding pair on the carbon atom is demonstrated by the Lewis-base behaviour of CO in the adduct $OC \cdot BH_3$. Carbon monoxide will also act as a ligand to many transition metals to give derivatives in which it is attached via the carbon atom to the metal, e.g. $Ni(CO)_4$, nickel carbonyl. For this reason CO is extremely poisonous because it complexes with the iron in haemoglobin in the blood and renders it incapable of picking up oxygen.

Carbon dioxide, CO_2, is much less reactive than CO. Its structure is linear and the bonding is simply explained as two double bonds, $O=C=O$. The $E(C=O)$ of these bonds is about 80 kJ mol^{-1} larger than that found in ketones, which is quoted in Table 12. Carbon dioxide dissolves in water to give a weakly acidic solution. Most of the CO_2 in such solutions exists as hydrated gas molecules $CO_2(H_2O)_x$ where x is unknown but probably represents a range of values. In such solutions there exist equilibria involving hydrogen carbonate and carbonate ions:

$$CO_2 \cdot H_2O \rightleftharpoons H_2CO_3 \rightleftharpoons H^+ + HCO_3^- \rightleftharpoons H^+ + CO_3^{2-}$$

The carbonate ion has a planar structure with all the C–O bonds the same length (129 pm) showing that the bonding is intermediate between a single and a double bond and the π-bonding is delocalized.

Silicon $(Ne)3s^2 3p^2$

Unlike carbon, silicon forms only one allotrope, with the diamond structure. In nearly all its derivatives it has a tetrahedral configuration at the silicon atom. It also has empty $3d$ orbitals but unlike phosphorus, sulphur, and chlorine it has no electrons to promote into these orbitals and so it cannot increase its valency beyond four. It can use these $3d$ orbitals if electrons are donated by another atom of a molecule or ion, and this explains why some silicon compounds exhibit Lewis-acid behaviour. Even so it is reluctant to act in this manner unless the energy of the bond which is formed is high and the donating atom is small in size. The most stable compound of this type is the anion SiF_6^{2-} which has an octahedral arrangement of fluorine atoms about the silicon.

Unlike carbon there are no double or triple bonds formed by silicon either to itself or to other second- or third-row elements. For this reason the chemistry of silicon is generally only a pale shadow of carbon chemistry except when silicon is bonded to oxygen. This combination produces a rich crop of compounds and structures.

SILICON–HYDROGEN   The compounds formed by silicon and hydrogen are called the silanes and like the alkanes have the general formula Si_nH_{2n+2}. However there is a limit to the value of n and the maximum seems to be about eight. The silanes are very reactive, especially towards oxygen and water; SiH_4 explodes in air. They can only be studied in vacuum apparatus like that developed by Stock for studying the boranes. The silanes are prepared by the action of an acid on magnesium silicide. A solution of ammonium bromide in liquid ammonia (which behaves as an acid solution − see p. 40) is the best for this purpose.

The differences between the alkanes and silanes can be anticipated by comparing the bond energies of Tables 12 and 13. Both the Si–H and Si–Si bonds are much

Table 13 Bond energies, E, and bond lengths, r, of silicon bonds

Bond	E/(kJ mol^{-1})	r/(pm)	Bond	E/(kJ mol^{-1})	r/(pm)
Si–H	323	146	Si–F	582	154
Si–O	485	151	Si–Cl	391	202
Si–Si	226	232	Si–Br	310	215
Si–C	301	189	Si–I	234	243

weaker than their carbon analogues. Most compounds containing a Si–H bond are very reactive and the silyl group, SiH_3^-, is rarely met in everyday chemistry, while the methyl group, CH_3^-, plays a very important role. Silicon-hydrogen compounds need special handling techniques such as vacuum systems and dry boxes. The electronegativity difference between carbon and silicon and their relativity to that of hydrogen has already been mentioned (p. 22).

SILICON–OXYGEN   Here the difference between carbon and silicon is very marked. The reason lies in the ability of carbon to form π-bonds with oxygen as in CO, CO_2, and CO_3^{2-}; a tetrahedral arrangement of singly bonded oxygen atoms around carbon is unknown. Silicon, on the other hand, does not form such π-bonds and the tetrahedral arrangement of oxygen atoms about silicon is the only one found in the silicon oxides and oxoanions (see Fig. 15).

In silicon dioxide itself each oxygen atom links together two silicon atoms and the resulting structure has little regularity. When liquid SiO_2 is cooled it produces glass, which is in effect a supercooled liquid. Only on long standing does this material crystallize, so that even Roman glassware which may be 2000 years old can still be seen to be only partially crystalline.

In the silicates the tetrahedral arrangement of oxygen atoms is maintained, but one, two, three, or even all four of these now in theory carry a negative charge. In the last case this gives the simple SiO_4^{4-} anion which is found in zircon, $ZrSiO_4$. If three oxygen atoms are charged one is left to bond to another silicon and the result is the $Si_2O_7^{6-}$ anion shown in Fig. 16(I). If two oxygens carry charges then two are free to bond

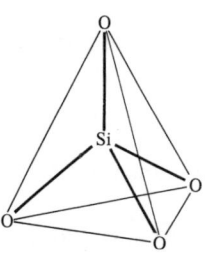 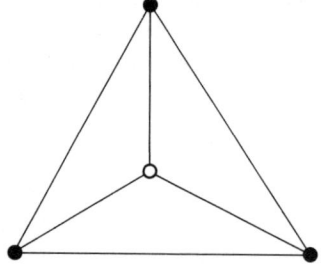

Fig. 15. SiO₄ unit and plan view

and this produces chains (II) or rings (III). The chainlike arrangement is found in MgSiO$_3$ and the ringlike structure in beryl, Be$_3$Al$_2$(Si$_6$O$_{18}$). When only one oxygen is charged, three can form bonds and this results in a layer lattice. Minerals with this structure betray their layer character by being easily cleaved into sheets. Mica and talc are of this type.

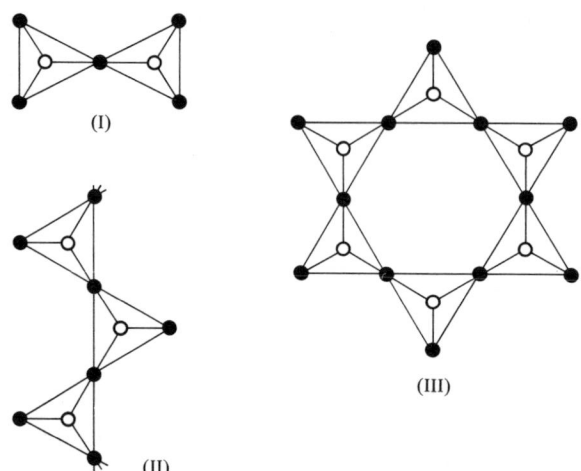

Fig. 16. The structures of the silicates (● = O, ○ = Si)

SILICON–HALOGEN   Comparison of bond energy values for halogen bonds to carbon and silicon can be made from Tables 12 and 13. In all cases the silicon-halogen bond is the stronger, yet surprisingly the silicon halides react rapidly with water whereas the carbon halides remain comparatively unaffected. The reason for this rests partly with the insolubility of the carbon halides in water but this is not the whole story. The Lewis acidity of the silicon compounds would encourage adduct formation with water as a first step to reaction. But even among the silicon halides the behaviour towards water varies. Silicon chloride, bromide, and iodide react to give silicon dioxide and the corresponding acid, e.g.:

$$SiCl_4 + 2H_2O = SiO_2 + 4HCl$$

Silicon fluoride also produces some silicon dioxide but the chief product is the hexafluorosilicate anion:

$$3SiF_4 + 2H_2O = 4H^+ + 2SiF_6^{2-} + SiO_2$$

In this reaction it is noteworthy that the number of Si–F bonds remains unchanged. (Compare in this respect the hydrolysis of BF_3 in the previous chapter.) Of all the silicon halogen bonds only that of the fluoride has a larger bond energy than $E(Si–O)$. Although the SiF_6^{2-} anion is known the other halides do not give similar anions. One of the reasons why $SiCl_6^{2-}$ does not form is that six large chlorine atoms cannot be packed around silicon without there being too great a repulsion between the chlorine atoms. In other words there would be *steric hindrance* in such an arrangement. Lower down group M4 where the elements are larger it becomes possible to get such anions as $SnCl_6^{2-}$ and $PbCl_6^{2-}$, because with these structures steric hindrance is much less.

SILICON–CARBON   Organosilicon compounds, as they are called, are a very important field of chemistry but are beyond the scope of this book. The best known are the methylsiloxanes, popularly called the silicones, which have a polymeric structure, and this confers upon them such useful properties as elasticity, water repellancy, and so on.

$$\begin{array}{c} CH_3 CH_3 CH_3 CH_3 \\ | | | | \\ -Si-O-Si-O-Si-O-Si-O- \\ | | | | \\ CH_3 CH_3 CH_3 CH_3 \end{array} \quad \text{polydimethylsiloxane}$$

7 Nitrogen and phosphorus

The elements of group M5 have the electronic configuration ns^2np^3 in the valence shell. Of the five electrons three are unpaired and available for bond formation and two are present as a non-bonding pair. The presence of the non-bonding pair is demonstrated by comparing the bond angle of methane and ammonia. Methane has the tetrahedral angle of 109·5° as expected, whereas the angle of ammonia is about 3° smaller as shown in Fig. 6. The reason for this is the electrostatic repulsion between the bonding and non-bonding pairs. The non-bonding pair of nitrogen and trivalent phosphorus compounds should also make their derivatives good Lewis bases, and this is often the case.

Nitrogen and phosphorus differ in several ways. The valency of phosphorus can be increased from three to five by making use of the empty $3d$ shell and promoting an electron of the non-binding pair: $3s^2 3p^3 3d^0 \rightarrow 3s^1 3p^3 3d^1$. Phosphorus is capable of a coordination number of six, although this is rare, but nitrogen is limited to a maximum c.n. of four. However nitrogen can form π-bonds to other second-row elements such as carbon and oxygen and form double and triple bonds, which phosphorus cannot form.

Nitrogen

Nitrogen, like carbon, is essential to life as we know it. On this planet nitrogen is found almost exclusively as the diatomic gas N_2 in the atmosphere. In this simple molecule there is a formal triple bond, N≡N, but unlike other triple bonds it is chemically unreactive. This poses problems since it is essential to convert it to compounds which living things can utilize. A natural process which does this is electric discharge (lightning) in the atmosphere which forms oxides of nitrogen and these are washed to earth as dilute nitric acid. Another, and more useful process takes place in certain bacteria, algae, and yeasts and this turns atmospheric nitrogen into ammonia. An important step in this process is the formation of a transition metal complex in which N_2 is primarily attached as a ligand. The search by inorganic chemists for these complexes has not yet been successful, but there have come to light other compounds which are capable of reacting at room temperature with N_2. The derivative di(isopropanoxy)titanium, Ti $(OCH(CH_3)_2)_2$, reacts with nitrogen gas to give $N_2Ti(OCH(CH_3)_2)_2$ and this can be reduced to ammonia, NH_3.

There are other processes for 'fixing' atmospheric nitrogen but these require high temperatures. The Haber process is the most widely used and this is based on the equilibrium of hydrogen and nitrogen. Relatively high yields of ammonia can be got by using pressures of the order of 10^7 Nm^{-2} at 800 K and an iron-alumina catalyst.

$$2N_2 + 3H_2 \rightleftharpoons 2NH_3$$

Table 14 gives the bond energies and bond lengths of the common nitrogen covalent bonds, and shows that the strongest single bond formed by nitrogen is that to hydrogen. This is reflected in the extensive chemistry associated with nitrogen-hydrogen compounds.

Table 14 Bond energies, E, and bond lengths, r, of nitrogen bonds

Bond	E/(kJ mol^{-1})	r/(pm)	Bond	E/(kJ mol^{-1})	r/(pm)
N–H	390	101	N–F	272	137
N–O	199	146	N–Cl	193	195
N=O	678	115	N–Br	116	214
N≡O*	1063	106	N–N	160	147
N–C	305	147	N=N	415	125
N=C	615	130	N≡N*	946	110
N≡C	891	116			

* bond dissociation energy, D

NITROGEN–HYDROGEN   The best known of these is ammonia NH$_3$ whose preparation and structure have already been mentioned. Despite having the strongest single bond, NH$_3$ is still a very reactive molecule. It will react with many non-metal halides, replacing the halogen with an amide group for example:

$$PCl_3 + 3NH_3 = P(NH_2)_3 + 3HCl$$

Ammonia is very soluble in water where it exists principally as NH$_3 \cdot$H$_2$O. This is in equilibrium:

$$NH_3 \cdot H_2O \rightleftharpoons NH_4^+ + OH^-$$

and removal of the hydroxide ion by the addition of acid causes the equilibrium to shift to the right and the solution to behave as if it were ammonium hydroxide.

Liquid ammonia has a b.p. of 240 K, which is much higher than expected due to extensive H-bonding in the liquid phase. Although its b.p. is below room temperature, evaporation can be kept to a minimum by the use of insulated flasks and it is quite easy to work with liquid ammonia as a solvent. There are three types of compound which are soluble in liquid ammonia. The first are organic compounds which contain oxygen or nitrogen atoms, such as alcohols, ROH, ethers, R$_2$O, and amines, RNH$_2$,

R_2NH, and R_3N. These, and many others, are capable of forming H-bonds with the solvent. The second class of compound is the ionic solids, and with these the factor controlling their solubility is the *anion*. This must be large and carry a single negative charge. Thus bromides, iodides, thiocyanates and nitrates are very soluble. Fluorides and hydroxides, though singly charged, are too small and are insoluble. Sulphates and phosphates, though large, are too highly charged and again are insoluble. The third class of compounds are the strongly electropositive metals and in particular the alkali metals.

Sodium will dissolve readily, without reacting, in liquid ammonia and can be reclaimed by evaporating the solvent. (Compare the effect of adding sodium to water!) The nature of these solutions is not fully understood but they behave as if sodium cations and free electrons were present. The solutions are conducting:

$$Na \rightleftharpoons Na^+ + e^-$$

and have excellent reducing properties as a result. They find wide use in organic chemistry.

Liquid ammonia, like water, undergoes a certain degree of self-ionization:

$$2NH_3 = NH_4^+ + NH_2^- \quad K = 10^{-33}$$
$$\text{cf. } 2H_2O = H_3O^+ + OH^- \quad K = 10^{-14}$$

and by analogy with water we can see the ammonium ion as the acid species and any compound which increases the concentration of this in liquid ammonia will behave as an acid in this solvent. Consequently ammonium bromide in liquid ammonia acts as an excellent acid. Similarly NH_2^- represents the base and a compound such as sodium amide, $NaNH_2$, in liquid ammonia acts as a base. The reaction of NH_4Br and $NaNH_2$ in liquid ammonia is an acid-base neutralization reaction for this solvent:

$$NH_4Br + NaNH_2 = NaBr + 2NH_3$$

Two other nitrogen-hydrogen compounds are known. These are hydrazine, N_2H_4, and hydrazoic acid, HN_3. Both are much less stable than ammonia and are explosive when pure.

NITROGEN–OXYGEN   There are six nitrogen oxides with the formulae N_2O, NO, NO_2, N_2O_4, N_2O_3, and N_2O_5. The last two are of minor interest. The first on the list, dinitrogen oxide, is a gas isoelectronic with carbon dioxide, CO_2, and has the same linear structure (O=N≡N) and low reactivity.

Nitrogen monoxide, NO, is also a gas but has the unusual feature of having an odd electron. Molecules with odd numbers of electrons are uncommon because the reason why atoms come together to form molecules is to pair off electrons. For this reason we would expect two molecules of NO to combine to give a dimer, N_2O_2, thereby pairing off the odd electrons. Nitrogen monoxide does not do this because the odd electron is in an antibonding orbital and is not free to form a covalent bond. The nitrogen-oxygen bond of NO is best thought of as a triple bond with a bond length of 110 pm and a bond energy of 627 kJ mol^{-1}. Removal of the odd electron gives NO$^+$ and now the length is 106 pm and the energy is 1063 kJ mol^{-1}, which is more in keeping with a triple bond. Nitrogen monoxide reacts rapidly with oxygen to produce nitrogen dioxide, another molecule with an odd number of electrons.

Nitrogen dioxide, unlike the monoxide, is capable of forming a dimer and the two forms exist in equilibrium:

$$2NO_2 \rightleftharpoons N_2O_4$$

In the solid and liquid phases the dimer, N_2O_4, predominates, but in the gas phase the amount of NO_2 increases as the temperature rises, and at 373 K there is 90% of this. The dimer has the structure:

$$\begin{array}{c} O \\ \diagdown \\ N \overline{175} N \\ \diagup \\ O O \end{array}$$

with a very long nitrogen-nitrogen bond, which is about 30 pm longer than the normal single bond. The reason for this is unknown.

Nitrogen dioxide reacts with water to give a mixture of nitrous and nitric acids; heating the resulting solution converts the nitrous acid to nitric and nitrogen monoxide is evolved:

$$2NO_2 + H_2O = HNO_2 + HNO_3$$
$$3HNO_2 = HNO_3 + 2NO + H_2O$$

Although nitrous acid is unstable its nitrite salts are stable. Nitric acid is stable and is a common oxidizing agent.

NITROGEN–HALOGEN   Nitrogen trifluoride is a stable gas obtained from the reaction of ammonia and fluorine. It is stable despite its rather low bond energy of 272 kJ mol^{-1}. It is thermodynamically unstable with respect to both N–H and N–O yet it is kinetically inert to hydrolysis. The molecule has a very low dipole moment of 0·780 x 10^{-30} C m which is much lower than that of ammonia at 4·88 x 10^{-30} C m. This may explain why NF$_3$ is 'unattractive' to other reagents.

Nitrogen trichloride, NCl$_3$, is extremely reactive and dangerous when in concentrated solutions. Dulong who first prepared it in 1811 from ammonia and chlorine lost three fingers when his sample exploded. The other halides NBr$_3$ and NI$_3$ are even more unstable, and the triiodide has never been obtained except as NI$_3$(NH$_3$)$_n$, where n can be from one to twelve.

Phosphorus

Phosphorus, like nitrogen, is essential to plant growth and animal life, but it is present on the surface of this planet as phosphate mineral deposits. The element can be obtained by heating phosphate rock with sand and coke in an electric furnace. Phosphorus can exist in several allotropic forms named after the colour, such as white, red, violet, or black phosphorus. White phosphorus consists of P_4 molecules which have a tetrahedral structure:

The bond energies and bond lengths of the main phosphorus covalent bonds are given in Table 15. The value of $E(P-O)$ can vary quite considerably and the value quoted in the table is that for the bond in P_4O_6. The values of P—F and P—Cl are for the trivalent phosphorus halides. Corresponding values for the pentavalent derivatives are slightly lower. (Why?)

Table 15 Bond energies, E, and bond lengths, r, of phosphorus bonds

Bond	$E/(\text{kJ mol}^{-1})$	$r/(\text{pm})$	Bond	$E/(\text{kJ mol}^{-1})$	$r/(\text{pm})$
P—H	322	142	P—F	490	154
P—O	407	164	P—Cl	317	204
P—P	209	221	P—Br	264	220
P—C	264	184	P—I	184	247

PHOSPHORUS–HYDROGEN   For nitrogen the bond to hydrogen is the strongest single bond. This, and the ability to form H-bonds, results in an extensive chemistry. For phosphorus the bond to hydrogen is weaker both absolutely and relatively, and there is no tendency to H-bond formation. Also the non-bonding electron-pair of PH_3 is more diffuse in space and this results in weaker Lewis-base properties.

Phosphine, PH_3, is a gas which can be prepared by the action of concentrated sodium hydroxide solution on white phosphorus. On heating such a mixture PH_3 is evolved and this burns on contact with air due to traces of diphosphine, P_2H_4, which is very unstable. Like ammonia, phosphine has a pyramidal structure.

PHOSPHORUS–OXYGEN   There are two oxides P_4O_6 and P_4O_{10}. Both have the same basic cage structure of four phosphorus atoms linked by oxygen atoms as shown in Fig. 17. The oxides are formed when phosphorus is burned in a deficiency (P_4O_6) or excess (P_4O_{10}) of oxygen. They react with water to give the two most important

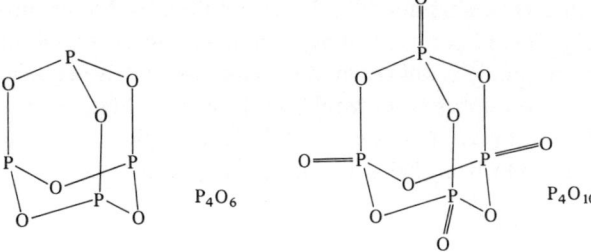

Fig. 17. The oxides of phosphorus

of the phosphorus oxoacids, phosphorous acid, H_3PO_3, and phosphoric acid, H_3PO_4.

The number of phosphorus acids is large but all have a coordination number of four at phosphorus. The three acids with only one phosphorus atom are illustrated in Fig. 18. The hydrogen atoms bonded directly to phosphorus are not acidic but the ones attached to oxygen are. Studies using D_2O solutions show that only the latter type of hydrogens exchange with the solvent.

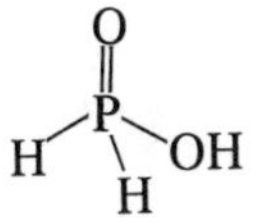

hypophosphorous 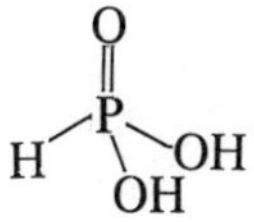   phosphorous 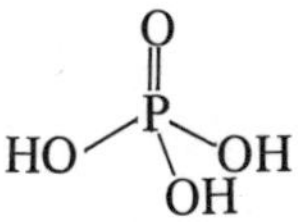   orthophosphoric

Fig. 18. The oxoacids of phosphorus

Like the silicates the phosphates have many and varied structures based on the number of negative charges and shared oxygen atoms attached to phosphorus. With one charge and two shared oxygen atoms the anions are chains or rings, called the metaphosphates. It is possible to get chains of polymeric phosphates which are several thousand units long.

PHOSPHORUS–HALOGEN   There is a large array of phosphorus-halogen compounds. Some are molecular and some are ionic, and some exist as an equilibrium mixture of the two forms. For example phosphorus pentachloride behaves in this manner in certain solvents:

$$2PCl_5 \rightleftharpoons PCl_4^+ + PCl_6^-$$

There are four trihalides, PF_3, PCl_3, PBr_3, and PI_3, with pyramidal structures as one would expect of compounds with one non-bonding electron-pair. They make good Lewis bases because the non-bonding pair is less diffuse in space than that of PH_3, and they readily form adducts with boron compounds, e.g. $Br_3B \cdot PCl_3$.

There are three pentahalides, PF_5, PCl_5, and PBr_5, and a few mixed pentahalides such as PF_3Cl_2. These have trigonal bipyramidal arrangements as illustrated in Fig. 6. Phosphorus pentaiodide is unknown but the packing of five large iodine atoms around a small phosphorus atom would result in steric hindrance. Also phosphorus pentabromide is not very stable and at 308 K it dissociates into phosphorus tribromide and bromine. Even PCl_5 at 433 K breaks up in the same way.

8 Oxygen, sulphur, and selenium

These non-metals have the outer electronic configuration ns^2np^4. They are two electrons short of the stable ns^2np^6 configuration and therefore it is not surprising to find the oxygen anion, O^{2-}, in many metal oxides. Outside the environment of the metal oxide lattice the oxide anion is less stable and in water, for example, it is rapidly hydrolyzed to OH^-. In the ns^2np^4 configuration there are two unpaired electrons and two non-bonding pairs and for oxygen this means a valency of two. Sulphur and selenium have vacant $3d$ and $4d$ orbitals into which an electron of a non-bonding pair can be promoted thereby increasing the valency from two to four and even to six.

$$ns^2np^4nd^0 \rightarrow ns^2np^3nd^1 \rightarrow ns^1np^3nd^2$$
$$\text{divalent} \quad \text{tetravalent} \quad \text{hexavalent}$$

The maximum coordination number of oxygen is four but this is found in very few compounds. A coordination number of three is fairly common as in such species as H_3O^+ and $(C_2H_5)_2O \cdot BF_3$ where it is behaving as a Lewis base. The maximum coordination number of sulphur and selenium is six.

Oxygen

This is the most abundant element of the earth's surface. It is found in the atmosphere as O_2, as water covering most of the planet's surface area, and combined with metals and non-metals in ores and rocks especially as silicates. The element exists in two allotropic forms — dioxygen, O_2, and ozone, O_3, both of which are gases. The former can be converted to the latter by passing it through a silent electric discharge.

Magnetic measurements show dioxygen to have two unpaired electrons. This can be demonstrated by suspending a small Dewar flask containing liquid oxygen near the poles of a powerful U-shaped magnet, such that the flask when empty can swing freely between the pole pieces. The presence of unpaired electrons in dioxygen proved a stumbling block for the valence bond theory which predicts a molecule with a double bond, O=O, and no unpaired electrons. Molecular orbital theory, on the other hand, can explain the presence of unpaired electrons by showing that these must occupy two equivalent anti-bonding orbitals. The bond dissociation energy of dioxygen is 498 kJ mol^{-1}. If one of the anti-bonding electrons is removed to give the dioxygen ion, O_2^+, this energy increases to 644 kJ mol^{-1}. (Compare NO and NO$^+$.) The ion O_2^{2+}, though yet unobserved, should have an even stronger bond.

Ozone has no unpaired electron but has a delocalized π-bonding system.

The bond length and bond energies of some covalent oxygen bonds are given in Table 16. Other bonds to oxygen are also to be found in tables in the other chapters. The strongest single bonds formed by oxygen, in order of decreasing bond strength, are B–O > S–O > Si–O > H–O > Se–O > P–O. All these bonds are very strong and it is therefore not surprising that these elements have well-developed chemistries of their oxides and oxoacids.

Table 16 Bond energies, E, and bond lengths, r, of oxygen bonds

Bond	E/(kJ mol^{-1})	r/(pm)	Bond	E/(kJ mol^{-1})	r/(pm)
O–H	467	96	O–O	138	149
O–C	336	142	O=O*	498	121
O=C	695	122	O–F	185	141
O≡C*	1074	113	O–Cl	207	168
			O–Br	201	178

* bond dissociation energy, D

OXYGEN–HYDROGEN   There are two hydrogen oxides — water, H_2O, and hydrogen peroxide, H_2O_2.

The chemistry of water is vast and that of its liquid phase is especially so since water is the solvent *par excellence*. The reason for this is partly due to its high dipole moment which allows it to solvate charged species such as ions very effectively, and partly due to its extensive H-bonding, which aids solution of many organic compounds containing oxygen and nitrogen atoms. In theory each water molecule is capable of forming four H-bonds but this is only achieved in ice at temperatures below 80 K (see Fig. 19). In the liquid phase there are two or three H-bonds per molecule and even near its b.p. there is still on average one H-bond per molecule.

Another feature of water is the small but very important self-ionization:

$$2H_2O \rightleftharpoons H_3O^+ + OH^- \quad K = 10^{-14}/298 \text{ K}$$
$$\quad\quad\quad\text{acid} \quad \text{base}$$

This is the basis of aqueous acid-base chemistry, since any compound which will increase the concentration of H_3O^+ will act as an acid, and anything which will increase the concentration of OH^- will behave as a base. An acid-base reaction then is the reverse of the equilibrium.

Hydrogen peroxide is also a good solvent and for the same reasons as water. When pure it is liable to explode, although diluted with water it is quite safe. Hydrogen peroxide is obtained by the electrolysis of sulphuric acid containing ammonium sulphate. The process involves the formation and hydrolysis of the peroxodisulphate ion and its subsequent hydrolysis. The hydrogen peroxide is removed by distillation at low pressures:

$$H_2S_2O_8 + 2H_2O = 2H_2SO_4 + H_2O_2$$

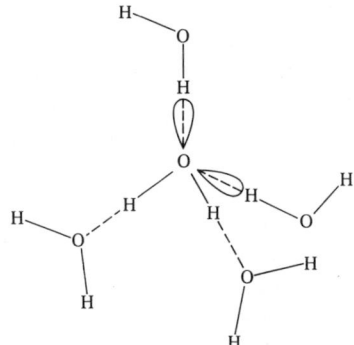

Fig. 19. Hydrogen bonding in water at low temperatures

OXYGEN–HALOGENS   There are two types of oxygen-halogen bond. There is the simple, straightforward, single covalent bond found in such compounds as oxygen difluoride, F_2O, dichlorine oxide, Cl_2O, and dibromine oxide, Br_2O. The bonds in these compounds are very weak as Table 16 shows and the compounds are accordingly very reactive; Br_2O is unstable with respect to bromine and oxygen at temperatures above 230 K. The second type of oxygen-halogen bond has a certain amount of double-bond character and is found in the other halogen oxide derivatives and in the halogen oxoacids. These compounds are discussed in the next chapter.

Sulphur and selenium

Because of the similarity between these two elements and their compounds they can be discussed together. Both have valencies of two, four, and six, in which there are two, one, and no non-bonding electron-pairs respectively. The bond lengths and bond energies shown in Table 17 are for the lowest valency state although there is not a great deal of difference between these and the higher valence derivatives.

Sulphur, like phosphorus, is known for its large number of allotropes. Some of these consist of S_6 and S_8 molecules with ring structures. Orthorhombic and monoclinic sulphurs have S_8 rings which are crown-shaped. Selenium is also found as Se_8

S_8

rings, and in addition there is a grey semimetallic form which used to be used in photoelectric cells because of its electron emission when exposed to light. The grey form is used in the manufacture of semiconductors. In this allotrope there are infinite chains of Se atoms in helices.

Table 17 Bond energies, E, and bond lengths, r, of sulphur and selenium bonds

Bond	E/(kJ mol^{-1})	r/(pm)	Bond	E/(kJ mol^{-1})	r/(pm)
S–H	347	133	Se–H	276	147
S–O	c. 500	150	Se–O	420	178
S–F	328	158	Se–F	285	181
S–Cl	255	199	Se–Cl	243	216
S–C	272	182	Se–C	243	198
S–S	226	208	Se–Se	172	232

When sulphur is heated the rings of S_8 open and join together to form long chain polymers. Gentle heating of sulphur powder in a tube containing a thermometer can be used to demonstrate the change which occurs above 400 K. Upon melting, a pale yellow mobile liquid is obtained which becomes more viscous as the temperature continues to rise due to the polymer chains growing longer. At the maximum viscosity the liquid can be poured into iced water to give threads of the allotrope known as plastic sulphur. On standing, this form eventually reverts to orthorhombic sulphur.

SULPHUR–HYDROGEN, SELENIUM–HYDROGEN Unlike O—H the bond to hydrogen is relatively unimportant in the chemistry of these elements. Hydrogen sulphide, H_2S, and selenide, H_2Se, are incapable of forming H-bonds and this markedly affects their physical properties; for this reason these derivatives are gases at room temperature. There is also a series of compounds called the sulphanes with the general formula HS_nH, where n can be from two to six sulphur atoms which are in a chain with the hydrogen atoms at either end. These compounds are unstable and disproportionate to H_2S and sulphur.

SULPHUR–OXYGEN, SELENIUM–OXYGEN When sulphur burns in air or oxygen it produces sulphur dioxide, SO_2, and a small amount of sulphur trioxide, SO_3. The trioxide is thermodynamically the more stable form but its formation in this reaction is slow. The reaction can be speeded up by the use of such catalysts as vanadium pentoxide.

Both SO_2 and SO_3 dissolve readily in water to give acid solutions. With SO_3 the product is sulphuric acid, H_2SO_4, but with SO_2 the principal species present in water is $SO_2(H_2O)_x$. This behaves as if it were sulphurous acid, H_2SO_3, because it is in equilibrium with hydroxonium ions:

$$SO_2(H_2O)_x \rightleftharpoons H_3O^+(H_2O)_{x-2} + HSO_3^-$$

Liquid sulphur dioxide, b.p. 263 K, is used as a solvent since it will dissolve a large number of salts and organic compounds. For salts to dissolve the anions must be large, e.g. bromides and iodides. Among the organic substances which dissolve are the aromatic hydrocarbons such as benzene, while the aliphatic hydrocarbons are insoluble. Sulphur dioxide was once used to remove aromatic compounds from paraffin and thereby reduce its smokiness when burnt. This process can be demonstrated in the laboratory by using liquid sulphur dioxide from a siphon to treat paraffin to which benzene has been added. The experiment should be carried out in a well-insulated flask to prevent loss of sulphur dioxide by evaporation, and in a ventilated fume cupboard.

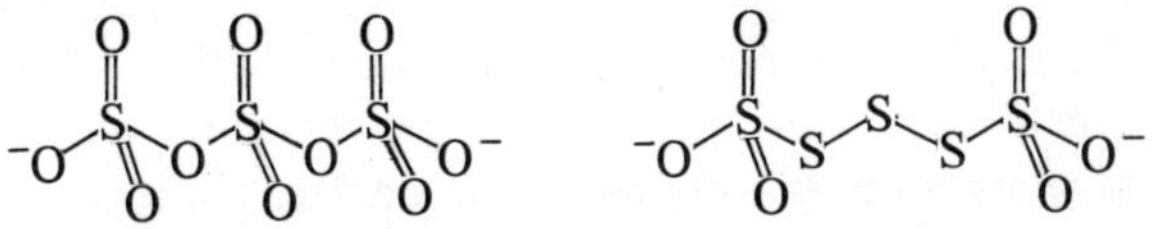

Fig. 20. The polysulphates (I) *and polythionates* (II)

There is no self-ionization in liquid sulphur dioxide of the type found in water or liquid ammonia, and so it is not possible to speak of acids and bases for this solvent.

In sulphuric acid and the sulphates there is a tetrahedral arrangement of oxygen atoms about sulphur, as was the case with the silicon and phosphorus oxoacids and salts. With the sulphates, however, there is not the variety of structures that were found with the silicates and phosphates. Apart from the simple sulphate ion there are chains of polysulphates (Fig. 20). As well as these chains there are the polythionates with the general formula $^-OS_3-S_n-SO_3^-$ which have chains of sulphur atoms, and n in the formula can be 0,1,2,3, etc. These are more examples of the catenation of sulphur.

SULPHUR–HALOGEN, SELENIUM–HALOGEN

Fluorides There are some divalent sulphur fluorides but these are not well characterized. The better known fluorides of sulphur and selenium are the tetrafluorides, SF_4 and SeF_4, and the hexafluorides, SF_6 and SeF_6. The structure of these compounds is illustrated in Fig. 6 in Chapter 3. Sulphur tetrafluoride is a very reactive gas which is easily hydrolyzed to HF and SO_2, and because of this it readily attacks glass. This in turn produces more water which attacks more SF_4, producing more HF which attacks yet more glass, and so on until all the SF_4 has been destroyed. Metal vessels which form an impervious metal fluoride coating, e.g. nickel, or polyfluoroethylene apparatus must be used when handling sulphur tetrafluoride.

Sulphur hexafluoride unlike the tetrafluoride is an inert gas which finds use as an insulator for high-voltage equipment. It will resist hydrolysis even at 700 K, and there are very few chemical reagents which react with it under normal conditions. The reason for the reactivity of SF_4 and the inertness of SF_6 does not lie in bond energy differences since in both compounds the bond energy is the same at 328 kJ mol^{-1}. The answer is the presence of the non-bonding pair of electrons of SF_4. This pair gives the molecule a dipole moment which thereby makes it 'attractive' to other chemical species with which it can react. The non-bonding pair also allows it to act as a Lewis base and with boron trifluoride an adduct $F_3B \cdot SF_4$ is formed.

Chlorides and *bromides*   The hexavalent state is encountered neither in the sulphur nor in the selenium chlorides nor bromides. Even the tetravalent state is uncommon and where it is found the compound is unstable, for example SCl_4 decomposes above 252 K. The divalent compounds SCl_2, S_nCl_2, and S_nBr_2 are known and in these there are chains of sulphur atoms with $n = 2,3$, etc. Corresponding selenium derivatives are known although here the chains are only short.

Iodides   There are no sulphur or selenium iodides. These combinations of elements seem incapable of forming stable bonds either as binary compounds or even in conjunction with a third type of element. This seems inexplicable as both oxygen and tellurium of group M6 are capable of forming covalent bonds with iodine. Tellurium forms a tetraiodide, TeI_4.

9 Fluorine, chlorine, bromine, and iodine

The halogens have the electronic configuration ns^2np^5 and are only one electron short of the stable ns^2np^6 shell. The gain of an electron to form the simple halide anion, X^-, plays a very important part in the chemistry of these elements. Many metal elements form ionic halides and the halide anions are stable in water.

The halogens resemble one another to a much greater extent than might have been expected from the other groups of this part of the periodic table. There are some features of fluorine chemistry which tend to set it apart from the other halogens, such as its low coordination numbers and the fact that it displays only monovalency. The maximum coordination of fluorine is two which it has when it is H-bonding or fluorine bridging. The maximum coordination numbers of chlorine and bromine are six, and of iodine eight; these are found in only a few compounds.

Because of the availability of empty d orbitals the valencies of chlorine, bromine, and iodine can be expanded from one to three to five, and in the case of iodine to seven. For iodine this involves the promotion of up to three electrons:

$$5s^2 5p^5 5d^0 \rightarrow 5s^2 5p^4 5d^1 \rightarrow 5s^2 5p^3 5d^2 \rightarrow 5s^1 5p^3 5d^3$$
$$\text{monovalent} \qquad \text{trivalent} \qquad \text{pentavalent} \qquad \text{heptavalent}$$

The elements

Fluorine is found chiefly as fluorspar, CaF_2, and cryolite, Na_3AlF_6. Electrolysis of liquid hydrogen fluoride containing potassium fluoride, KF, as electrolyte is used to generate the gas F_2. This is the most reactive molecular species known and will react with all the elements except helium, neon, and argon. This reactivity partly stems from the low bond energy of F_2 which, as Table 18 shows, is lower than that of chlorine or bromine. The F—F bond is weakened by the coulombic repulsions of the three non-bonding electron-pairs on each fluorine atom.

Chlorine occurs as NaCl in vast deposits of rock salt, and as the principal solute in sea water. Bromine is also a constituent of sea water and is extracted from it. Iodine is much rarer, and although it too is present in the sea no economic process exists for its profitable extraction from this source. Certain oil-well brines are rich in iodine, and in Chile iodate deposits are found. Like fluorine the other halogens exist as diatomic molecules in the elemental state.

Derivatives of the halogens

Much of the chemistry of these elements has been dealt with in previous chapters. Bond energies and bond angles for many of the non-metal to halogen bonds can be found in the tables associated with these chapters. Table 18 gives these values for the bonds which the halogens form between themselves.

Inspection of all the bond energy tables show that for a particular element the bond strengths to the halogens are in the order:

$$M-F > M-Cl > M-Br > M-I$$

Also as a general rule it appears that the greater the electronegative difference between the atoms comprising the bond the stronger is the bond. The reader is invited to test this observation by extracting the appropriate data from the bond energy tables. Where the rule breaks down in the third row of the periodic table the electronegativities of the elements concerned could be responsible and the reader is invited to calculate revised electronegativity values for these elements.

Table 18 Bond energies, E, and bond lengths, r, of halogen-halogen bonds

X	Fluorine E/(kJ mol^{-1})	r/(pm)	Chlorine E	r	Bromine E	r	Iodine E	r
F–X	153	144	172	163	197	176	247	257
Cl–X			243	200	219	214	210	232
Br–X					196	229	178	250
I–X							152	267

HALOGEN–HYDROGEN Hydrogen fluoride is a liquid, b.p. 292 K, and as such is used as a solvent. Many organic compounds are soluble in it and electrolysis of their solutions converts C–H bonds to C–F bonds. This process is used to manufacture the fluorocarbon compounds used as refrigerants and aerosol propellants. The high b.p. of liquid hydrogen fluoride is due to strong H-bonds. In the solid and liquid states there are zig-zag chains of HF molecules held together by H-bonding:

$$\cdots F-H\cdots F\diagdown_{270}^{H}\cdots F-H\cdots F\diagdown H\diagup^{140°}_{F}\diagup H\cdots F\diagdown H\cdots F\diagup$$

The only stable salts in liquid HF are fluorides and these dissolve to give the bifluoride anion, HF_2^-, which has the symmetrical structure:

$$[F\text{---}H\text{---}F]^-$$
$$\longleftarrow\!\!226\!\!\longrightarrow$$

There is a certain amount of self-ionization in liquid HF:

$$3HF \rightleftharpoons \underset{\text{acid}}{H_2F^+} + \underset{\text{base}}{HF_2^-}$$

and it is possible to speak of acids and bases for this solvent. Fluorides can be seen as bases since they increase the concentration of the bifluoride ion.

Hydrogen chloride, bromide, and iodide boil at temperatures well below that of hydrogen fluoride because with these compounds there is virtually no H-bonding. All dissolve readily in water to give very strong acids. However, in dilute aqueous solutions, HF is a weak acid (about as strong as acetic acid), but at high concentrations it is a strong acid. (Why?)

HALOGEN–HALOGEN Derivatives formed from the halogens themselves are called interhalogen compounds. Fluorine in particular is capable of forming derivatives with the other halogens and often induces them to show their maximum valency and coordination number. The known halogen fluorides are:

ClF	BrF[1]	(IF)[2]	[1]Disproportionates to BrF_3 and Br_2.
ClF_3	BrF_3	IF_3[3]	[2]Unknown.
ClF_5	BrF_5	IF_5	[3]Stable only below 238 K.
		IF_7	

Chlorine monofluoride is the only stable diatomic interhalogen compound. It can be prepared from a mixture of fluorine and chlorine at 520 K. At 540 K the main product is chlorine trifluoride, which has a T-shaped structure as shown in Table 6 in Chapter 3. Bromine trifluoride has the same configuration.

All the halogens form a pentafluoride, but chlorine pentafluoride requires very high pressure and temperature for its production. All have the umbrella-shaped molecule which is explicable by the Sidgwick-Powell rules for six electron-pairs, one of which is a non-bonding pair (Table 6).

Only iodine is capable of forming a heptafluoride, IF_7, and this has a pentagonal bipyramid structure. It is formed from iodine and excess fluorine at 520 K.

In addition to the halogen fluorides there are mixed halogen compounds not involving fluorine such as ICl, which can easily be prepared in the laboratory by passing chlorine over iodine. If the system is under pressure the formation of ICl_3 can be demonstrated.

HALOGEN–OXYGEN The oxygen dihalide compounds have been discussed in the previous chapter. Fluorine has only the one oxide, F_2O, but the other halogens have several oxides as well as the monoxide. Chlorine has an oxide ClO_2, chlorine

dioxide, which is a molecule with an odd number of electrons. The bond energy E(Cl–O) in this molecule is 255 kJ mol^{-1}, which is 43 kJ mol^{-1} higher than that of the Cl–O bond of OCl_2. The bond length is also 20 pm shorter, and together these facts suggest a delocalized bonding arrangement in chlorine dioxide. There are two other chlorine oxides — unstable Cl_2O_6 and stable Cl_2O_7 (b.p. 353 K), but nothing is known of their structures.

Bromine oxides are known and these resemble the chlorine oxides. The iodine oxides however are radically different. They are all solids and are thought to have ionic structures. The most stable is I_2O_5 which has the structure $IO^+IO_4^-$. The others are I_2O_4 ($IO^+IO_3^-$ or $I^+IO_4^-$) and I_4O_9 ($I^{3+}(IO_3^-)_3$) which is unstable.

In water the halogen oxides give the oxoacids, except for oxygen difluoride, there being no known fluorine oxoacid. The simplest acids are the hypochlorous, hypobromous, and hypoiodous acids of the general formula HOX. These tend to decompose readily and their salts are even less stable. The only acid with two oxygen atoms attached to the halogen atom is chlorous acid, HOClO. Its salts, the chlorites, are fairly stable.

The halic acids, $HOClO_2$, and their salts the halates are much better known. The potassium salts, $KClO_3$, $KBrO_3$, and KIO_3 are easily prepared and their separation from the corresponding halides is a useful exercise in fractional crystallization. Simple tests can be carried out on the products, such as the effect of heat, and the reactions of their aqueous solutions (acidified with dilute sulphuric acid) and sodium nitrite.

The perhalic acids, $HOXO_3$, and the perhalate anions, XO_4^-, are known for chlorine and iodine. Although the oxygens are limited to a tetrahedral arrangement about chlorine, with iodine both tetrahedral and octahedral structures are found. Complex periodates can be produced by the sharing of one or two oxygen atoms of the octahedral form to give complex oxoanions as were found with the silicates, phosphates, and sulphates.

At the start of this chapter it was noted that the halogens tend to accept an electron to give the simple X^- anion. It therefore comes as something of a surprise to find that iodine can also form cations I^+ and I^{3+}, involving the loss of one or three electrons. The iodine oxides are thought to have these cations in some of their forms. When iodine is dissolved in pure sulphuric acid these cations are also formed and colour the solution blue (I^+) or brown (I^{3+}). They can be stabilized by acting as Lewis acids for very strong Lewis bases. For example, the strong base pyridine, C_5H_5N, gives the species $[C_5H_5NI]^+$ which is even stable in organic solvents. The electrolysis of such solutions produces iodine at the cathode showing that this was present in the solution in a cationic form.

10 The rare gases

The rare gases are helium, neon, argon, krypton, xenon, and radon. All have the stable outer electron configuration $ns^2 np^6$, except that helium has only two electrons and the $1s^2$ configuration but this is also unreactive. Where vacant d orbitals exist these can be utilized in the formation of covalent bonds as with krypton and xenon. Argon might also have been expected to use its $3d$ orbitals as phosphorus, sulphur, and chlorine do, but so far no argon derivatives have been reported. Krypton appears willing to promote only one electron, $4s^2 4p^6 4d^0 \rightarrow 4s^2 4p^5 4d^1$, as in KrF_2. Only xenon seems capable of multiple valency and when one speaks of the chemistry of the rare gases one is generally referring to xenon and its compounds. Radon is extremely rare and radioactive in all its isotopes. No doubt its chemistry, were it to be investigated, would resemble that of xenon.

The rare gases are present in the atmosphere from which they can be obtained by the fractional distillation of liquid air. Helium is also found in large quantities in some American natural-gas wells. Argon is the most abundant rare gas comprising 1% of the atmosphere. This is the result of the radioactive decay of one of the naturally occurring potassium isotopes, ^{40}K, which decays by electron capture to give ^{40}Ar. Radon also occurs naturally as one of the products of radioactive decay of thorium and uranium, but in quantities too small to collect. The source of radon used in laboratories is from sealed ampoules of radium which decays to give radon gas.

Xenon

Before the discovery of xenon derivatives in 1963 the only compounds that the rare gases seemed capable of forming were clathrate compounds. In these the gas atoms are trapped in large holes in the crystal lattices of hydroquinone, and are not bonded in the normal sense of the word. The first xenon compound was made by Bartlett. He had prepared the new compound $O_2 PtF_6$ which is ionic and of the form $O_2^+ PtF_6^-$. The ionization potential of the oxygen molecule is almost the same as that of xenon and he reasoned that a corresponding xenon derivative might also exist. The result was that he prepared $XePtF_6$ and this was the key to the door of xenon chemistry.

The ionization potential of xenon can be measured using a xenon-filled thyratron tube. The energy of cathode emitted electrons of such a thyratron is increased to a level that causes ionization of the gas. A small current (< 1 μA) due to Xe^+ appears at the target electrode which is kept at a large negative potential. A graph of applied voltage against observed current is plotted and from this the ionization potential of xenon can be deduced by extrapolation. For details of this experiment see Leaflet 7 of the Mullard Educational Electronic Experiments Series.

About twenty-five xenon compounds are now known. The most stable are the fluoride, oxoanions, and oxides in that order. In its ground state xenon has the outer electron configuration $5s^2 5p^6 5d^0$. Promotion of one, two, three, and even four electrons is possible to give a variety of derivatives as Table 19 shows.

Table 19 Some compounds of xenon

Electron configuration	$5s^2 5p^5 5d^1$	$5s^2 5p^4 5d^2$	$5s^2 5p^3 5d^3$	$5s^1 5p^3 5d^4$
Valency	2	4	6	8
Compound (m.p./K)	XeF_2 (413)	XeF_4 (387)	XeF_6 (319)	XeO_4
			$XeOF_4$ (301)	$Na_4 XeO_6$
			XeO_3	
			$Na_2 XeO_4$	
			$Na_2 XeF_8$	

The fluorides are made from mixtures of xenon and fluorine gases. Passing them through an electric discharge gives XeF_2; passing them through a heated tube gives XeF_4; or heating them under pressure gives XeF_6. Hydrolysis of the fluorides with controlled amounts of water produces the oxygen derivatives. Hydrolysis with sodium hydroxide solution produces the salt $Na_2 XeO_4$ from XeF_4, and $Na_4 XeO_6$ from XeF_6. The oxoacids $H_2 XeO_4$ and $H_2 XeO_6$ are unknown, although when XeO_3 is dissolved in dilute NaOH the sodium hydrogen salt of xenic acid is produced.

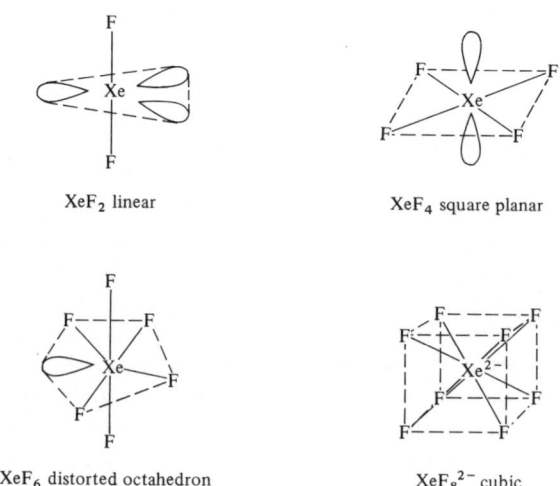

Fig. 21. The structures of the xenon fluorides

The structures of the xenon fluorides, XeF_2, XeF_4, XeF_6, and XeF_8^{2-} are explicable on the assumption that they have three, two, one, and no non-bonding electron-pairs respectively (see Fig. 21). The reader might like to predict the structures of XeO_3, XeO_4, XeO_4^{2-}, and XeO_6^{4-}.

The bond energies of the xenon and krypton compounds which have been measured are given in Table 20 and as expected they are very weak. Using these values it was possible to calculate the electronegativities of krypton and xenon given in Table 7 of Chapter 2.

Table 20 Bond energies, E, and bond lengths, r, of krypton and xenon bonds

Bond	$E/(kJ\ mol^{-1})$	$r/(pm)$
Xe–F	133	195
Xe–O	88	180
Kr–F	50	188

Since xenon hexafluoride has a non-bonding electron-pair it might be expected to display Lewis-base behaviour. With BF_3 an adduct $XeF_6 \cdot BF_3$ is formed, but it has been suggested that this compound is ionic and has the structure $XeF_5^+BF_4^-$ rather than a xenon-boron bond, which in any case would be very weak.

Bibliography

Introductory texts

HUDSON, M. *Energy and Bonding.* EUP, 1969
SPICE, J. E. *Chemical Binding and Structure.* Pergamon, 1964
WALLACE, H. G. and COWELL, V. W. *Chemistry: an Electronic and Structural Approach.* Murray, 1970
TAYLOR, R. J. *Water.* Unilever Educational Booklet – Advanced Series No. 5, 1966
COE, J. S. *Chemical Equilibrium.* Methuen, 1971.
DAWSON, B. E. *Energy in Chemistry.* Methuen, 1971.

More advanced texts

DASENT, W. E. *Inorganic Energetics.* Penguin, 1970
JOLLY, W. J. *The Chemistry of the Non-metals.* Prentice-Hall, 1966
HESLOP, R. B. *Numerical Aspects of Inorganic Chemistry.* Elsevier, 1970
COTTON, F. A. and WILKINSON, G. *Advanced Inorganic Chemistry.* Interscience, 1966
HARVEY, K. B. and PORTER, G. B. *Physical Inorganic Chemistry.* Addison-Wesley, 1963
SANDERSON, R. T. *Inorganic Chemistry.* Reinhold, 1967

Experimental chemistry texts

DAWSON, B. E. *Practical Inorganic Chemistry.* Methuen, 1967 (2nd Edn.)
FOWLES, G. *Lecture Experiments in Chemistry.* Bell & Sons, 1963 (6th Edn.)
Nuffield Chemistry Publications: Laboratory Investigations. Longman/Penguin

Reference works

COTTRELL, T. L. *The Strength of Chemical Bonds.* Butterworths, 1954
Interatomic Distances. Special Publication No. 11, Chemical Society, 1958

Index

Acid-base reactions, 23, 40, 46, 47, 53
Acids, 23
Allotropes, 31, 42, 45, 47
Allred-Rochow electronegativity, 14
Ammonia, 39
Antibonding orbitals, 41, 45
Argon, 55
Astatine, 1

Bartlett, 55
Benzene, 7
Beryl, 36
Bifluoride ion, 52
Boiling points of the elements, 2
Boiling points of the hydrides, 23
Bond angles, 18
Bond dissociation energy, 9
Bond energy, 10
Bond length, 8
Bonding electron-pairs, 5
Boranes, 28
Borates, 30
Boric acid, 30
Boric oxide, 30
Borohydrides, 29
Boron, 26
Boron trifluoride, 17, 29
Bromine, 51
Bromine fluorides, 53
Bromine oxides, 47, 53

Carbon, 31
Carbon dioxide, 18, 34
Carbon monoxide, 33
Carbonates, 18, 34
Catenation, 33, 48, 49
Chlorate, 54
Chlorine, 51
Chlorine fluorides, 53
Chlorine oxides, 47, 53
Clathrates of the noble gases, 55
Conjugate acid-base pair, 23
Coordination number (c.n.), 16
Covalent bond, 5
Covalent radii, 8

d Orbitals, 31
Delocalization, 6
Deuterium, 24
Diamond, 32
Diborane, 23, 28
Diboron tetrachloride, 11
Dioxygen, 45
Diphosphine, 12, 42
Dipole moment, 13
Donor π bond, 29, 33

Electrical conductivity of the elements, 2
Electron affinity, 14
Electronegativity, 12
Electron-pair repulsion, 18

Fluorides, 3
Fluorine, 51

Glass crystallization, 35
Graphite, 32

H-Bonding, 22
H-Bridging, 22, 28
Haber process, 38
Halogens, 51
Helium, 55
Haemoglobin, 34
Hexafluorosilicate, 37
Hydrazine, 40
Hydrazoic acid, 40
Hydrofluoric acid, 52
Hydrogen bonding, 22
Hydrogen bridging, 22, 28
Hydrogen peroxide, 47
Hydrogen sulphide, 48
Hypohalous acids, 54
Hypophosphorous acid, 43

Ice, 46
Interhalogens, 53

59

Interstitial hydrides, 21
Iodates, 54
Iodic acid, 54
Iodine, 51
Iodine fluorides, 53
Iodine oxides, 54
Ionization potential, 14, 55
Isoelectronic compounds, 29, 33, 40

Krypton, 55
Krypton difluoride, 55

Lewis acids, bases, and adducts, 26
Lithium hydride, 21
Lone pairs, 18

Magnesium silicide, 35
Mean bond energy, 10
Melting points of the elements, 2
Metalloids, 1
Metaphosphates, 43
Mica, 36
Molecular structure, 16
Mulliken's electronegativity, 14

Neon, 55
Nitrite ion, 6
Nitric acid, 41
Nitrogen, 38
Nitrogen fixation, 38
Nitrogen fluoride, 41
Nitrogen oxides, 40
Nitrous acid, 41
Non-bonding electron-pairs, 18

Odd electron molecules, 40, 41
Orthophosphoric acid, 43
Oxoacids – see appropriate element
Oxygen, 45
Ozone, 46

Paramagnetism of oxygen, 45
Pauling electronegativity, 13
Perchloric acid, 54
Periodic table of the elements, viii
Phosphates, 43
Phosphoric acids, 43
Phosphorous acid, 43

Phosphorus, 42
Phosphorus halides, 43
Phosphorus oxides, 42
Pi bonds, 5
Potential energy of a diatomic molecule, 6
Proton, 20

Radon, 1, 55
Rare gases, 55
Repulsions of electron-pairs, 18

Selenium, 47
Selenium halides, 49
Selenium oxides and oxoacids, 48
Sidgwick-Powell theory, 17
Sigma bonds, 5
Silanes, 35
Silicates, 36
Silicon halides, 36
Silicon oxides, 35
Silicones, 37
Silyl group, 35
Single bond, 5
Sodium in liquid ammonia, 40
Stereochemistry, 16
Steric hindrance, 37, 44
Stock, 28, 35
Structure of molecules, 16
Sulphanes, 48
Sulphates, 49
Sulphur, 47
Sulphur dioxide solvent, 48
Sulphur fluorides, 50
Sulphur trioxide, 48
Sulphuric acid, 49

Talc, 36
Tellurium iodide, 50
Thionates, 49
Tritium, 24

Valence bond theory, 17

Water, 24

Xenon, 55
Xenon fluorides, 56
Xenon oxides and oxoacids, 56

Zircon, 35